Engineering Plastics and Their Commercial Development

Engineering Plastics and Their Commercial Development

A symposium co-sponsored by the Division of Chemical Marketing and Economics and the Division of Polymer Chemistry at the 157th Meeting of the American Chemical Society, Minneapolis, Minn., April 15-16, 1969

George F. Foy

Symposium Chairman

ADVANCES IN CHEMISTRY SERIES **96**

AMERICAN CHEMICAL SOCIETY

WASHINGTON, D. C. 1969

Coden: ADCSHA

Library of Congress Catalog Card 78-102896

PRINTED IN THE UNITED STATES OF AMERICA

Advances in Chemistry Series

Robert F. Gould, *Editor*

AMERICAN CHEMICAL SOCIETY PUBLICATIONS

FOREWORD

ADVANCES IN CHEMISTRY SERIES was founded in 1949 by the American Chemical Society as an outlet for symposia and collections of data in special areas of topical interest that could not be accommodated in the Society's journals. It provides a medium for symposia that would otherwise be fragmented, their papers distributed among several journals or not published at all. Papers are refereed critically according to ACS editorial standards and receive the careful attention and processing characteristic of ACS publications. Papers published in ADVANCES IN CHEMISTRY SERIES are original contributions not published elsewhere in whole or major part and include reports of research as well as reviews since symposia may embrace both types of presentation.

CONTENTS

PREFACE

Commercial development can be defined as the overall operation which combines, directs, and coordinates the functions necessary for guiding a new product from applied research to full commercialization. Included in these functions are such activities as use research, marketing research, process development, and market development.

Because the commercial development activity, particularly in the field of plastics, is so frequently misunderstood, or little understood, this symposium was undertaken with an aim toward clarifying the activities involved in developing a new plastic to a profitable commercial product. This symposium was presented on a "how-to-do-it" basis and is composed of a series of papers by experts in each of the fields and functions covered. Each author is eminently qualified because of his years of successful experience in this particular area of expertise. Prices referred to were those in effect in early 1969.

The general background for the symposium and the scope of the coverage is presented in the first paper. It is, however, probably desirable to present certain definitions at this point to further set the stage for the papers which follow:

(1) *Use Research.*, Use research is applied research in the laboratory, field, or plant aimed at determining the potential profitable applications for the new plastic. Such research would include the evaluation and comparison of the new material with potentially competitive material now successfully used.

(2) *Market Research.* According to Reinhold's "Chemical Marketing Research," edited by N. H. Giragosian, marketing research is "that function directed toward determining the need, acceptance, potential volume, expected competition, possible selling price, most suitable packaging, use patterns, and future trend for a product. It means providing data on the sales potentials and estimates of potential profits of a given product. Marketing research embraces surveys of the location of the market and specifies acceptable qualities for possible new products."

(3) *Process Development.* Process development is defined according to Wiley's "Successful Commercial Chemical Development," edited by H. M. Corley, as "a combination of applied research and chemical engineering research for the purpose of determining the most economically desirable and technologically

feasible method for producing either new or already established products."

(4) *Processing Development.* This is the development of an optimum processing system for a new plastic through the use and possible modifications of fabricating concepts and equipment.

(5) *Market Development.* As Defined in Wiley's "Successful Commercial Chemical Development," edited by H. M. Corley, market development is "that field of promotional effort between basic research and regular sales, by which existing or potential markets for new products and entirely new fields of utility for old products are developed and tested for profitable salability." Included in this function is joint development activity between the producer of the product and the potential user, the distribution of samples and sales literature, packaging, labeling and pricing, among other functions.

It is hoped that this symposium is a worthwhile contribution to our industry. In addition to acknowledging the excellent work and fine cooperation of the authors of each of the papers, special acknowledgement is also made to Richard Germann of Abbott Laboratories and John Dockum of PPG Industries for their assistance in organizing and executing this symposium.

GEORGE F. FOY

Montclair, N.J.
June, 1969

1

Summary of Engineering Plastics

GEORGE F. FOY

Chemical Business Development Co., Montclair, N. J. 07042

Engineering plastics are those which possess physical properties enabling them to perform for prolonged use in structural applications, over a wide temperature range, under mechanical stress, and in difficult chemical and physical environments. In comparison with metals for a given application, plastics may offer such advantages as transparency, self-lubrication, economy in fabricating and decorating. Plastics can be flexible, they are electrical nonconductors and thermo insulators, with these properties being either advantages or disadvant ages, depending on use. Properties of plastics can be modified through the use of reinforcing agents, fillers, and chemical additives. Engineering applications for plastics include mechanical units under stress, low friction components, heat and chemical resistant units, electrical parts, housings, high light transmission applications, building construction functions, and many miscellaneous uses.

The functional steps necessary for the commercial development of a new engineering plastic are much the same, as far as definitions are concerned, as those required for developing a new chemical. These functional steps can be defined as:

1. USE SCREENING
2. USE RESEARCH
3. MARKETING RESEARCH
4. PATENT–LEGAL ASPECTS
5. PROCESS AND PROCESSING DEVELOPMENT
6. MARKET DEVELOPMENT
7. SELLING AND SERVICING

The above functional operations are in a general chronological order but there is much overlap with respect to time and function, and there is a continued interplay between the people executing each activity. Each of these commercial development functions are discussed in detail by experts in the following chapters. However, this paper will prepare the way by outlining the general background and giving a birds-eye view of the industry.

As additional brackground, Thermoplastics, Thermosets, and the Thermoplastic-Thermoset materials are discussed along with their engineering properties and applications. And because of the somewhat different nature of the products, a paper covering the use of mill shapes for engineering purposes is included, along with a paper discussing engineering plastics from the users standpoint.

Let's embark on our symposium now by taking a look at the rise of the plastics to engineering applications.

Early Engineering Materials

Prior to the Civil War, engineers were pretty well limited to working with only a few materials, such as iron, copper, tin, zinc, brass and bronze and, of course, wood. Steel became more readily available shortly after the Civil War through the wider use of the recently discovered Bessemer and open-hearth processes and the subsequent discovery of alloy steels. And around the turn of the century, aluminum was becoming available in commercial quantities.

Arrival and Growth of Plastics

Although cellulose nitrate, a semisynthetic plastic and not really an engineering material, was introduced in 1868, the first completely synthetic plastic, phenol-formaldehyde, was introduced in 1909. Phenol-formaldehyde is certainly an engineering material and the first of many such products to excite the imagination of engineers.

Since the introduction of phenol-formaldehyde in 1909, a steady stream of new plastics has been appearing on the scene. Today there must be close to 50 separate families of these materials, not including alloys or other modifications of the basic types. Not all of these have engineering applications, but most do.

The significance of this history is that over a span of many centuries engineers had only about nine basic materials with which to work. Further, they were limited in their design considerations by the properties of these few materials. So, within just the past 60 years we have seen the advent of these fifty or so separate families of plastics and literarily hundreds of modifications of these materials.

Highlighting some of these post-phenolic developments, we saw polyvinyl chloride introduced in 1927, acrylics in 1936, nylon in 1938, and fluorocarbon in 1943. ABS (acrylonitrile-butadiene-styrene) was introduced in 1948, acetal in 1956, polycarbonate in 1957, polyphenylene oxide in 1964, and polysulfone in 1965.

More recently, we have seen other new plastics arrive, including such exotic, high performance materials as the polybenzimidazoles, polyoxadiazoles, polyperfluorotriazines, polyphenylenes, and such inorganic materials as the boron polymers, the metalloxanes, and the polysilazanes, to name only a few. These exotic materials are mostly development products today, they are very expensive and will undoubtedly find their first uses, if any, in the aerospace industry which requires the high performance offered by these materials and can afford to pay for them. Some of these new materials will become commercial successes, others will not.

The growth of all plastics in general, and engineering plastics in particular, has more than kept pace with the general progress in technology. Most importantly, the demands of the hot wars in recent years and, of course, the cold war have stimulated the development of new and better plastics. Peacetime requirements, insofar as they can be separated today from war needs, have also stimulated the development and use of new plastics. Consider, for example, the growing use of plastics in the transportation, appliance, electro-electronics, and other industries. Indeed, such needs and developments have in recent years brought into being an entirely new discipline, namely that of materials engineering.

The materials engineer has been most important in stimulating and accelerating the use of engineering plastics. The activities of identifying his company's needs with respect to applications, standards, and costs, feeding this information to suppliers, and working with the suppliers' development people to obtain the desired material is funneled through the materials engineer. This is a tremendous development advantage, particularly when working with the larger companies.

Definition and Types of Engineering Plastics

Up to now we have been speaking about engineering plastics in a very general way but, before going further, let's define an engineering plastic. No single definition will satisfy everyone, but for the purpose of this discussion we will use the following: Engineering plastics are those which possess physical properties enabling them to perform for prolonged use in structural applications, over a wide temperature range, under mechanical stress, and in difficult chemical and physical environments.

Engineering plastics are most frequently thought of as the acetals, nylons, fluorocarbons, phenolics, polycarbonate, and polyphenylene oxide, to name just a few. These are indeed engineering materials and for such applications are usually used in relatively small amounts, in comparison with the nonengineering plastics which are used in commodity quantities. They are materials characterized by high performance properties and, with the exception of phenolics, are generally priced at from around 50 cents to over $1.00 per pound. However, if one is willing to stretch the definition for an engineering plastic, at least from some points of view, other materials, such as the stronger and more durable polyolefins, ABS (acrylonitrile-butadiene-styrene), the acrylics, rigid polyvinyl chloride, the urethane elastomers, and some other materials also find engineering uses.

Advantages and Disadvantages of Plastics

For most engineering applications, plastics are considered to be competitive primarily with metals, although there is, of course, considerable competition among the plastics themselves. In comparison with metals, plastics have certain properties which are generally considered to be advantageous for engineering applications. For the most part, plastics have better chemical and moisture resistance and are more resistant to shock and vibration than metals. They are lighter in weight and usually either transparent or at least translucent in thin sections. They have the advantage of absorbing sound and vibration and some possess greater wear and abrasion resistance than metals. Some of them, as in the case of nylons, for example, are self-lubricating. Significantly, one of the most important characteristics of plastics is that they usually are easier to fabricate than metals. As you know, some plastics can be plated but perhaps an even more important property is that plastics can be pigmented in a wide variety of colors. Finally, because of their lighter weight, giving many of them an advantage in cost per cubic inch with respect to metals and, because they are usually easier to fabricate, finished parts made of plastics are frequently less costly than those made of metal.

In continuing this comparison of plastics with metals, we must recognize that plastics are characterized also by certain negative properties. To begin with, plastics are not as strong as metals. Generally speaking, they possess lower heat resistance and most of them are flammable. They are characterized by a much larger thermoexpansion, are often less ductile than metals, and most are much more subject to embrittlement at low working temperatures. And, unfortunately, for many applications for which plastics are used or desired to be used, they are more susceptible to creep. Plastics are softer than metals and some of them, when they absorb water or solvents, will change in dimension, quite a negative feature when these materials are used for gears and other close tolerance parts. In addition, most plastics are subject to degradation by ultraviolet light. Finally, most plastics cost more than metals on a per-pound basis and some of them cost more than some metals on a per-cubic-inch basis.

When plastics are compared with metals on the usual advantage *vs.* disadvantage consideration, some of their properties can be considered either favorable or unfavorable, depending upon the application. Among these are the softness and flexibility characteristics of plastics, even considering the more rigid types. Unfilled plastics are thermal and electrical non-conductors and they are subject to deformation by heat and/or pressure. Fortunately, however, both for the engineer and the plastics manufacturer, plastics can be modified by the addition of other materials. Pigmentation, for example, has already been referred to. In addition, fire retardants, ultraviolet absorbers, and other additives can be incorporated into plastics to improve their properties, and fillers and reinforcing materials can be incorporated into the plastics to impart many advantages.

Uses of Fillers, Reinforcing Agents, and Chemical Additives

It was previously stated that some of the polyolefins were used for certain engineering applications. The incorporation of glass fibers, for example, into a polyolefin will increase its strength, its toughness, its rigidity, its dimensional stability, and even its heat resistance, thus putting the reinforced material into an entirely different class than that material which has not been so modified. Other reinforcing agents for plastics include asbestos, which increases strength and heat resistance; and carbon black, which increases weather resistance and imparts lubricity and electrical conductivity. The addition of metal powders or metal fibers provide thermo and electrical conductivity to plastics, increase the strength of the material, and impart other desirable properties. Even materials such as talc and wood flour, which might be more properly called fillers, offer advantages, such as improved dimensional stability, improved impact strength and, importantly, lower cost.

What we are trying to emphasize here is that plastics of themselves are quite functional materials but they can sometimes be made more so through the addition of fillers, reinforcing agents, and chemical additives. Perhaps the most important contribution of fillers is the reduction in cost which they impart. The more expensive reinforcing agents, on the other hand, are used because of the desirable properties they impart to the plastic and such reinforced plastics will cost up to 30% more and even greater than the unreinforced materials.

Engineering Uses of Plastics

Let's take a look now at some of the engineering uses of plastics. To begin with, these uses fall into several general categories and we might best consider such applications from this standpoint.

Mechanical Units Under Stress. Cams, gears, couplings and such other mechanical components are examples of these applications. Requirements of a plastic for such uses would include high impact and high tensile strength, excellent stability and resistance to fatigue, along with the ability to perform for long periods at high temperatures. Good environmental resistance is sometimes required for these applications and the materials must be easily formed and perform continuously to close tolerance. The plastics most frequently used for such applications include the acetals, nylons, fabric-filled phenolics, and polycarbonates. The principal competitive materials for these uses are brass, iron, and steel.

Low Friction Components. Wear surfaces, slides, bearings and guides are components requiring low friction qualities. The most important requirement for these applications is, of course, a low coefficient of friction. Also important are good dimensional stability, heat resistance, and good abrasion resistance. The plastics most used for low friction applications include the fluorocarbons, nylons, acetals, and even ultra-high molecular weight polyethylene. Competitive non-plastic materials for these applications include the Babbitt metals, bronze, iron, and graphite.

Chemical and Heat Resistant Equipment. Chemical and other processing equipment, certain under-the-hood automotive parts and aerospace components are examples of these applications. Obviously the most desired properties of plastics going to such uses would include the excellent resistance to high temperatures and corrosive environments, along with good strength and, in some instances, resistance to shock and vibration. Because of the unique demand placed on such requirements, the highly stable plastics are used here, such as the fluorocarbons, chlorinated polyether, and glass-reinforced epoxy. The expensive metals, such as the stainless steels and titanium are the materials most competitive with plastics for these applications.

Electrical Parts. Example applications here are connectors, relays and other electrical and electronic parts. Good electrical resistance, of course, is an important required property. Other required properties are good dimensional stability, good tensile strength, and impact resistance. Candidates for these applications are mostly the thermosets, such as the alkyds, aminoplastics, epoxies, and phenolics, but certain thermoplastics are also used here including polycarbonate and polyphenylene oxide. Glass and ceramic would, of course, come to mind as principal materials competitive with plastics for these electrical uses.

Containers, Ducts and Housings. Plastics going to these applications should have good strength in all respects and good environmental resistance. They should be easily formed, with relatively low cost being a prime factor. Plastics finding wide use here include ABS, polypropylene, glass-polyester combinations, cellulose acetate butyrate, and some others. Competitive materials for these applications include steel, aluminum, and die-cast metals.

Glazing – High Light Transmission Units. Applications here include lenses and protective domes, in addition to glazing. Obviously, high light transmission and even transparency are required properties for plastics going to these applications. Other desired properties of plastics for these uses include good impact strength and frequently a high resistance to ultraviolet radiation. Products mostly used here are, of course, acrylics which possess excellent ultraviolet resistance, and polycarbonate which has unusually high impact resistance. Other plastics used here include polystyrene, cellulose acetate, and rigid vinyl. Glass, of course, is the principal competitive material.

Miscelianeous Applications. There are many such uses, some of which will be questioned as engineering. Some of these applications include building construction, luggage, safety-helmets, seating, kitchen utensils, rope and strapping, shoe parts, rollers, furniture, and even dentures. This list can go on almost indefinitely and the plastics used, reflecting desired properties, will vary widely.

It is probably safe to say that most engineering plastics are formed by molding but many other forming techniques are used, including machining, sintering, stamping, and casting. Other forming techniques are, of course, used also.

Plastics for the engineering applications we have just discussed will find some use in practically any industry you can name. You have all seen such lists in the trade literature but, again, for a general background here are a few such user areas: aerospace and military, appliances, business machines and other electrical-electronics equipment, communication equipment, laboratory and engineering instruments and equipment, plumbing, transportation, and such general catch-all areas as special machinery and products going to the recreation industry. Again, we could continue almost indefinitely and include such use industries as adhesives for structural applications, and even packaging where strapping and high-performance containers are required. These are all engineering applications, even though there is a tremendous gray area about which there has been and will continue to be much argument. But regardless as to whether the plastic goes into a black area or gray area, uses, volumes, and numbers will continue to grow as long as man's needs and his imagination continue.

RECEIVED June 6, 1969.

2

Engineering Thermoplastics from a Commercial Development Viewpoint

RUDOLPH D. DEANIN

Lowell Technological Institute, Lowell, Mass. 01854

*Typical engineering thermoplastics include ultra-high-molecular-weight polyethylene, poly(4-methylpentene-1), poly-*p*-xylylene, chlorinated polyvinyl chloride, fluoroplastics, acetal resins, chlorinated polyether, polyphenylene oxide, polysulfone, polycarbonate, nylons, polyimides, and glass-fiber-reinforced thermoplastics. These most frequently offer advantages in processability, ultra-thin-film formation, rigidity, strength, impact-resistance, lubricity, abrasion-resistance, high-temperature strength and stability, dimensional stability, dielectric properties, clarity, chemical resistance, water resistance, and especially in an improved balance between combinations of these properties. Major applications are largely as mechanical parts in machinery and applicances; electrical insulation in electrical and electronic instruments; and tanks, pipes, fittings, and gasketing in process industry equipment. They are processed primarily by injection molding, occasionally by extrusion, and often even by machining. New materials will continue to join the list at a fairly steady rate, and a few may eventually grow to commodity status.*

Since their commercial appearance a century ago, plastics have grown at an accelerating rate, and will probably exceed metals in volume by 1985 (Figure 1). Their growth rate greatly exceeds conventional materials such as metals, ceramics, wood, rubber, textiles, and paper, largely because plastics frequently offer superiorities in processability, flexibility, strength/weight ratio, impact strength (greater than ceramics, textiles, and paper), range from lubricity to adhesion, abrasion-resistance, energy absorption of

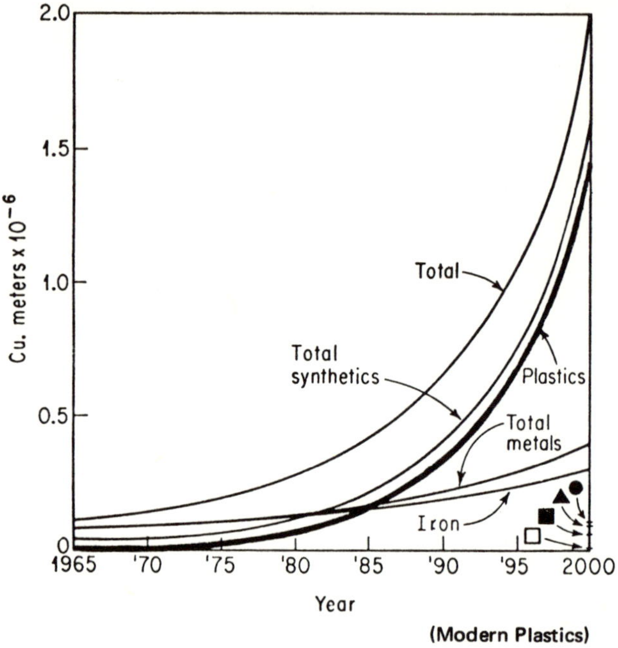

Figure 1. Annual consumption of metals, synthetics, natural fibers, and rubber, by volume (2) "Reproduced by permission, Modern Plastics Magazine, Mc Graw Hill, Inc."

● *Aluminum* ■ *Copper and zinc*
▲ *Synthetic rubbers and man-made fibers*
□ *Natural rubber and fibers*

foams, thermal and electrical insulation, range of color and clarity, and resistance to inorganic chemical corrosion(1).

At the same time, plastics suffer from a number of deficiencies which limit their acceptance and growth. Processability is still not easy or versatile enough for all applications. Rigidity is generally lower than metals, ceramics, and wood. Strength is generally lower than metals and ceramics. Brittleness is greater than metals and wood. Low hardness permits scratching and marring by abrasion of inorganic materials. Coefficient of thermal expansion is much higher than metals and ceramics, causing difficulty in matching components over a wide temperature range. Plastics become much more brittle at low temperatures, and soften sooner than metals, ceramics, and wood at high temperatures. They also suffer thermal decomposition and oxidation at high temperatures, and are more flammable than metals and ceramics. Outdoor weathering causes gradual and often fairly rapid deterioration. Resistance to organic chemicals is often poor. In some plastics absorption of moisture causes severe instability of dimensions and properties. Permeability is much higher than metals and glass, causing problems in packaging. And price is often higher than competitive materials.

For all these reasons, organic chemists and chemical engineers expend considerable ingenuity developing new plastic materials to remedy these deficiencies, and such new materials have been appearing commercially at a fairly steady rate for the past half century. Thermoplastics have enjoyed the greatest growth, because of their economical synthesis from low-cost raw materials and their adaptability to mass production by continuous processing techniques.

Engineering Thermoplastics

In the most general sense, all plastics are engineering materials, in that they offer specific properties which we judge quantitatively in the design of end-use applications. Among the large-volume established thermoplastics, we should certainly pay tribute to the engineering performance of the polyolefins, polystyrene, impact styrene, ABS, vinyls, acrylic, and cellulosic plastics.

Generally, however, when plastics marketers speak of engineering plastics, they use the term to distinguish the newer materials which are still used in smaller volume and at higher price, and often are still seeking their optimum markets and applications on the basis of specific superior properties which will justify their higher cost. Properties most often improved include processability, rigidity, strength, impact-resistance, lubricity, abrasion resistance, high-temperature strength and stability, dimensional stability, dielectric properties, clarity, chemical resistance, water resistance, and especially an improved balance between combinations of these properties. These improvements most often lead to applications as mechanical parts in machinery and appliances; electrical insulation in electrical and electronic instrumentation; and tanks, pipes, fittings, and gasketing in process industry equipment.

In such a flexible definition of engineering thermoplastics, the exact list and number of materials and/or families of materials will vary with the viewpoint of the specialist who prepares the list. For purposes of illustration, let us consider 13 thermoplastics, developed during the past decade or two, which are produced in small to moderate volume at medium to high prices, and are finding growing application in products where their superior properties justify their higher costs. Most of these descriptions of engineering thermoplastics are taken from the manufacturers' bulletins and unpublished discussions with manufacturers and users of these materials.

Polyolefins. ULTRA-HIGH-MOLECULAR-WEIGHT POLYETHYLENE. This is easily made by conventional low-pressure coordination polymerization. Hercules Hi-fax 1900 has a molecular weight of 2.5-5.0 million. Its outstanding properties are low coefficient of friction (0.11); abrasion-resistance superior to nylon, polyurethane, and steel; unbreakable in the Izod notched impact test; and high resistance to most inorganic and many organic chemicals.

Its greatest deficiency is processability—it is hardly thermoplastic at all; Hercules sells mill shapes which must be machined into end products, while Formica has developed a continuous compaction-sintering machine for producing it in sheet and laminate form (3). In addition, its modulus, strength, and heat distortion temperature are no better than conventional medium-density polyethylene. Bar, plate, and rod stock sell for $2 up, before machining.

It is finding industrial machinery applications, based mainly on its high lubricity and abrasion resistance, such as timing screws on bottling lines, suction box covers for fourdrinier screens in paper-making, textile loom pickers, coal mining chutes, wear strips on brewery bottling and canning lines, and sporting goods such as runners for snow skis. Large-scale growth will probably depend upon development of new low-cost processing methods.

POLY(4-METHYLPENTENE-1). It was first discovered by Natta, later researched by many companies for synthetic fiber, and finally commercialized by ICI as a specialty molding resin. It is mainly isotactic, 40-65% microcrystalline, and has the lowest density of any plastic, 0.83, which may be approaching the theoretical minimum.

It also has a high melting point, 240°C.; excellent electrical properties; and higher clarity than any other polyolefin. Its weaknesses are low rigidity and solvent resistance; sensitivity to oxidation; and high permeability, which can sometimes be converted into a virtue.

At $1.25/pound it is finding applications in hospital and laboratory ware in this country; earlier marketing in Great Britain has developed additional applications in lighting fixtures, milking machine and liquor dispenser sight glasses, sink traps, slot enclosures for electrical motors, and packages for reheatable foods. Further specialty applications are developing gradually, but large-scale growth would depend upon the much lower potential price inherent in the propylene-based starting material.

POLY (P-XYLYLENE). It was discovered by Szwarc, researched by several companies, and finally commercialized experimentally by Union Carbide. It is prepared by pyrolyzing *p*-xylene at high temperature and vacuum, then condensing on a cool surface, and is thus primarily adapted to the production of uniform thin films and coatings.

It has excellent dielectric properties, between Mylar and Teflon; extreme solvent and water resistance; and very low permeability. While it has a crystalline melting point of 400°C., it degrades at 200-300°C. in inert atmosphere and oxidizes at 60-100°C. in air, so it is not a high-temperature material. Vacuum deposition is a difficult custom operation, which costs about $250/pound; the high melting point and insolubility prevent any conventional thermoplastic processing techniques.

It is of interest primarily for very uniform ultra-thin films and coatings (0.002-5 mils) in applications such as electrical resistors, thermistors, thermocouples, stator cores, connectors, fast-sensing probes, photo cells, memory units, dropwise steam condensers for recovery of sea water, pellicles for beam splitters in optical instruments, windows for nuclear radiation counters, panels for micrometeorite detection, dielectric supports for planar capacitors, encapsulation of reactive powders, and supports in x-ray and optical work. Any significant growth would depend upon a major breakthrough in process techniques and a consequent lowering in price.

Polyvinyl Halides. CHLORINATED POLYVINYL CHLORIDE. It was produced in Germany up to three decades ago, but this was primarily a 1,1-disubstituted product of increased solubility for dry-spinning of fibers. Goodrich has developed a light-activated suspension chlorination process which produces 1,2-dichlorinated structures of increased hot strength, thermal stability, and flame resistance.

This Hi-temp Geon is superior to conventional rigid PVC primarily in its higher heat distortion temperature (208-234°F. at 264 p.s.i.). On the other hand, processing is somewhat more difficult, and cost of chlorination brings the price to $0.50. Primary applications are in residential hot and cold water tubing, hot water piping, and hot chemical process equipment such as pipe, plating baths, hot acid fume exhaust from steel pickling, spray etching, and metal finishing.

FLUOROPOLYMERS. These form one of our oldest and most spectacular families of engineering plastics. Polytetrafluoroethylene was developed by DuPont over two decades ago, and more recently by Allied Chemical, Hoechst, ICI, Pennwalt, and other manufacturers as well. It combines unusually low adhesion and friction, high temperature and flame resistance, excellent electrical properties, and extreme chemical inertness. Its high melting point and melt viscosity make thermoplastic processing extremely difficult, so that many products are more readily made by machining from stock shapes. It also suffers from low modulus and strength.

Its balance of properties is so unusual that even at $3.25/pound and specific gravity 2.2 it finds growing use in a wide variety of specialized applications ranging from electrical wire insulation in motors, locomotives, aircraft, missiles, spacecraft, lighting fixtures, stoves, ovens, switches, controls, and computers; electrical insulators for radar and television; and chemical pipes, fittings, valves, and pumps, hydraulic and fuel hose in aircraft, trucks, buses, and trains; to gaskets, packings, bearings, and cooking utensils.

More recently, modified fluoroplastics such as fluorinated ethylene/propylene copolymer, polychlorotrifluoroethylene, and polyvinylidene fluoride have been offered by DuPont, Allied Chemical, 3M, and Pennwalt respectively, to provide improved processability and mechanical strength at some sacrifice in heat-resistance, electrical properties, and chemical resistance; and at prices of $3.70-7.15 these have also been finding appropriate if smaller markets.

Polyethers. ACETAL RESINS. These stabilized polyoxymethylenes were introduced dramatically by DuPont and Celanese as engineering plastics to replace non-ferrous metals. Good mechanical strength, resilience, fatigue-resistance, lubricity, abrasion-resistance, heat distortion temperature, water and solvent-resistance can approach the behavior of metals on a volume basis, while processability, color possibilities, and corrosion-resistance are superior. Major weakness is sensitivity to thermal, oxidative, and ionic degradation.

At $0.65 they are developing growing markets in autos, appliances, plumbing, and hardware, such as gears, bearings, switch housings, valves, fan blades, razors, office equipment, pumps, conveyor chain links, and handles. While growth has not been as phenomenal as originally hoped, last year's tonnage of 45.5 million pounds was certainly handsome for a 9-year-old specialty plastic.

POLY (3,3-BIS(CHLOROMETHYL)OXETANE) This is marketed by Hercules as Penton chlorinated polyether. Its thermal, flame, and chemical resistance are used primarily for corrosion-resistant equipment in the process industries, such as valves, fittings, pumps, meters, and linings for steel pipe and tanks, for service in many corrosive atmospheres up to 250°F. or higher. At $4.50/pound, growth possibilities appear limited by increasing competition from lower-cost materials.

POLYPHENYLENE OXIDE. This is actually poly (2,6-dimethylphenylene oxide) and was introduced by General Electric as the first commercial representative of the new technique of oxidative polymerization. It offered a combination of mechanical strength, creep resistance, dimensional stability, good constant electrical properties, moisture resistance, and exceptionally high heat distortion temperature (345°F.); its major limitations were difficult thermoplastic processing, low solvent resistance, and sensitivity to thermal oxidation.

It was then modified by polyblending with impact styrene to produce Noryl, with good thermoplastic processability and somewhat lower heat distortion temperature (265°F.). In this form at $0.59/pound, it has been finding growing acceptance in business machines, appliances, electrical equipment, and water distribution equipment.

POLYSULFONE PLASTICS. These plastics which were commercialized by Union Carbide are actually aromatic polyethers containing periodic sulfone groups which provide additional resonance stabilization. They have good mechanical properties, creep resistance, and dimensional stability; but their outstanding quality is their high heat distortion temperature (345°F.) and resistance to thermal oxidative degradation. Limitations are difficult thermoplastic processability, amber color, and sensitivity to organic solvents.

At $1/pound they are finding applications in electrical and electronic equipment, hot household appliances, automotive under-the-hood parts, and aircraft ducts and panelling, as well as hot water service and metal/metal adhesives.

Polyester. BISPHENOL A POLYCARBONATE. This was developed simultaneously by Bayer and General Electric and represented a twin breakthrough: it was the first commercial application of interfacial polycondensation, and it demonstrated that the organic carbonate linkage was surprisingly stable in an aromatic high polymer. The resulting polymer has an unusual combination of high impact strength (12-17.5 f.p.i. for a 1/8 inch bar), heat distortion temperature (270°F.), and clarity but has poor resistance to alkali and many organic solvents.

Originally in a class by itself as a clear, impact and heat resistant thermoplastic, its growth has been slowed considerably by the onrush of more recent engineering thermoplastics. Nevertheless, at $0.80/pound, it reached 27 million pounds last year, going largely into lighting, appliances, electrical and electronic equipment; and continued growth prospects appear quite bright.

Polyamides and Imides. NYLONS. These probably were the first of the engineering thermoplastics, featuring a new combination of easy processability, mechanical strength, lubricity, abrasion resistance, hot strength, electrical resistance, and chemical resistance. A by-product of the synthetic fiber industry, they began replacing metals in applications such as mechanical components in automobiles and machinery, electrical equipment, and molded appliance parts. Competition from the newer engineering thermoplastics, and regeneration of waste fiber, has brought the price of standard nylon molding resins down to $0.75/pound, but they have retained their markets and continued to grow, reaching 73 million pounds last year.

The most serious technical limitation of nylons 6 and 66, their absorption of atmospheric moisture with consequent instability of dimensions and properties, has been overcome somewhat by the successive development of nylons 610, 11, and 12, offering lower moisture absorption and softer but stabler mechanical properties at prices of $1.20-1.60, and going into applications such as battery cases, pumps, timer gears, power tools, hydraulic and gasoline hose, gasoline containers, food packaging film, and hot melt adhesives for laminated textiles. Even considering the growing competition, the future of nylons as engineering plastics appears assured.

POLYIMIDES. They were developed first by DuPont and Monsanto for ultra-high-temperature applications such as electrical insulation, machine bearings, and structural components in instrumentation and aerospace development. These are theoretically linear but not at all processable by thermoplastics techniques.

American Cyanamid has been developing a thermoplastic polyimide XPI with somewhat lower temperature possibilities (heat distortion temperature 440°F.), which may open a new range of potential applications for such materials. Certainly the recent appearance of pots and pans with an external coating of polyimide, for decorative appearance and easy cleaning, represents a dramatic new breakthrough in the prejudice against using plastics in high-temperature applications.

Reinforced Thermoplastics

Short-Fiber-Reinforced Thermoplastics. They began with the use of short glass fibers to increase the modulus of polystyrene 14 years ago, and have grown to the use of 10-40% by weight of short glass fibers in nearly all the commercial thermoplastics. Such reinforcement combines some of the conventional benefits of glass fiber reinforcement in general, with most of the advantages of thermoplastic processability.

Properties most often improved are modulus, strength, and dimensional stability, approaching the properties of die-cast metals.

Other properties which are improved in some systems include low-temperature impact strength and heat distortion temperature. Led by Fiberfil and Liquid Nitrogen Processing, many of the original resin manufacturers have now also begun to offer similar materials, and applications are growing lustily into a wide variety of engineering and high-performance products.

Future Growth

These are typical of the present engineering thermoplastics. Others which could have been discussed include poly-1-butene among the polyolefins; polyvinylidene chloride and polyvinyl butyral among the vinyls; the newer ABS, vinyl, and acrylic polyblends; linear phenoxy resins; and saturated linear polyesters. Most of the engineering thermoplastics will continue to grow into larger volume and broader applications, with corresponding decrease in cost. A few will drop by the wayside. And new ones will continue to appear at a fairly steady rate, offering continued improvements in specific properties, and especially in balance of critical properties for specific applications. Most probable areas for major growth are in block copolymers, in sophisticated composites—both semi-compatible polyblends and reinforced plastics—and in newer and easier processing techniques for high-temperature polymers.

Selection of Optimum Materials and Application

With this continued growth in variety of plastics materials, selection and matching of optimum combinations between materials and applications becomes a major problem. With at least 50 types of commercial plastics already available, each in a variety of copolymers, molecular weights, and different manufacturers, there are already too many for rational manual choice of the optimum material for any specific application; and the situation is growing worse at an alarming rate.

To avoid complete chaos, we badly need to convert to computerized searching in the very near future. Such a search system would start first with the absolute property requirements, P_i, and select only those materials which passed this first screening test. Second, it would balance the relative importance f_i of other properties p_i, preferably on the basis of cost per unit property. And third it would present the design engineer with the one or several materials whose balance of properties ΣP would be best suited to his needs.

$$\Sigma P = P_1 P_2 P_3 \ldots P_i \ldots P_n [f_1 p_1 + f_2 p_2 + f_3 p_3 + \ldots f_1 p_i + \ldots f_n p_n]$$

Materials manufacturers should issue computer cards instead of brochures. Large users of materials could run their own computers, while small ones might best unite to support one master computer run by SPE or SPI for their benefit.

Product Design Theory

In the ultimate pairing of materials and applications, small short-run decisions are generally based on the particular manufacturer's immediate position: a materials manufacturer will seek any conceivable use for his products, a processor will try to process any possible material into any possible end-product, and a manufacturer of consumer products will use any material and process which opens the market fastest.

But for large-scale markets with a long life projection, the selection must be much more objective and impartial. To make a new end product, the designer should first draw up preliminary designs which indicate the general properties that will be required, then use computer searching to identify the one or several materials most likely to meet his needs, then return to his product and examine optimum designs using each of these materials, then consider the process techniques required for each material⇌design combination, and finally refine this 3-way choice to make his final decision.

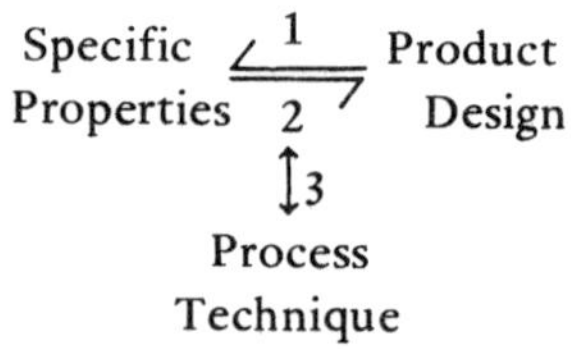

Beyond this initial theoretical drawing-board-and-slide-rule approach, of course, lies the critical need for experimental proof in actual process machinery and prototype field trials, which should be included as a 4th dimension in such a schematic diagram, and is the final pragmatic judgment to determine technical success of any plastic product. This is the direction in which applications research and product design must grow in the future, in order to convert plastics art into plastic science.

Literature Cited

(1) Jacolow, Melvin F., managing ed., "Modern Plastics Encyclopedia," McGraw-Hill, New York, 1968-69.

(2) "Modern Plastics Guide to Marketing," p. 1, McGraw-Hill, New York, 1966-67.
(3) Power, G. E., Meyers, K.D., Fremont, R. E., *U. S. Patent* **3,286,008**, (1966).
RECEIVED June 6, 1969.

3

Thermosets

ROBERT W. BAINBRIDGE

Durez Division, Hooker Chemical Corp., Walck Road, N. Tonawanda, N. Y. 14120

Thermoset molding compounds are classified as: alkyd, diallyl phthalate, epoxy, melamine, phenolic, polyester, silicone, and urea. Each material has its own particular characteristic. These materials are found in the following market areas: appliance, automotive, closure, communication, electrical switch gear, electrical starter gear, home laundry, power tool, textile, and many other allied products. The materials are all processed by performing a molding operation requiring heat and pressure. A range of molding equipment is available. The molding operation may be done either semi-automatically or fully automatic. Injection molding of thermosets is now possible and many parts are molded in this manner. In the past, some applications have been converted to thermoplastic materials; however, with new processing techniques and resin systems this trend may be reversed.

Thermoset molding materials are those plastic compounds that when molded form a permanent shape when heat and pressure are applied to them, while confined within a hardened steel mold.

The most common and widely used thermoset molding compounds are classified as follows: (a) alkyd, (b) allylic (diallyl phthalate), (c) amino (melamine and urea), (d) epoxy, (e) phenolic, (f) polyester, and (g) silicone. There may be other specialty thermoset resin materials used on specific applications.

Being man-made, thermoset molding compounds may be processed, using various fillers, to give most any property desired in an end product. The basic resin itself will impart certain desirable results when compounded with fillers, which will produce an end product that may have superior electrical, physical, and heat resistant characteristics, etc.

The widespread and growing use of these man-made materials can be credited in large part to their unique combinations of advantages. The advantages of thermoset materials are light weight, range of color, good physical properties, excellent electrical values, heat resistance, chemical resistance, dimensional stability, adaptability to mass production methods and often at lower cost. Recent development in processing permits extremely short cycles; therefore extremely high production at low tooling costs.

As stated, the basic resin system makes it possible to process a recommended molding compound for a specific end use or application. The general physical, electrical, specific properties, and their applications, may more clearly be understood if each type thermoset molding compound is examined.

Alkyd. Alkyd resins are the reaction product of a polyfunctional alcohol and a polyfunctional acid. When the resin is combined with a filler, catalyst, pigment, and release agent the resultant product is a granular or putty or rope-type molding compound. Fillers used may be fine mineral, glass, mineral and nylon—depending on the end product desired.

Alkyd molding compounds, when molded into finished parts, may have excellent arc resistance (180 sec. minimum) and excellent arc track resistance. Other good electricals such as dielectric strength and insulation resistance can be obtained. The filler combinations may also result in excellent dimensional stability and outstanding retention of electrical and physical properties after exposure to elevated temperatures (450°F.) for long periods of time. Water absorption values are extremely low, which helps to maintain electrical values.

Some formulations produce parts that may have Underwriters' Laboratories rating of 155°C. – 170°C. and 180°C. Formulations are also designed to meet the requirements of MIL-M-14F specifications. Granular-type compounds may be processed having impact strengths approaching 1 ft. lb. per inch of notch. Putty or rope-type compounds may be produced having further improved impact strength.

Alkyd molding compounds produce parts which are quite light stable and may be formulated in a wide range of color. Black, red, and natural are the most popular colors.

Typical applications for alkyd molding compounds are: small, delicate, high strength television tuners and tube bases, and potting cases. In the field of switch gear and motor control, alkyd compounds are used for their excellent arc track resistance, dimensional stability, and high physical strengths—particularly the high tensile and flexural characteristics. Automotive ignition system probably is the largest volume user of alkyd compounds. The excellent arc track resistance makes an ideal material for coil tops, distributor caps, and rotors.

Figure 1: Automotive circuit breaker (switch gear)

The alkyd molding compounds, depending on the form of the material, may be molded on a wide range of conventional molding equipment. Because of the short time required to cure the compounds, extremely fast molding cycles are possible.

Raw material costs may range from 32¢ to 75¢ per lb.

Diallyl Phthalate. Allylics are in another group of thermosetting molding compounds. The most widely used are the diallyl phthalate compounds. In general, most diallyl phthalate molding compounds are sold for military applications. Molded parts produced from these compounds have excellent electrical characteristics—especially insulation resistance at elevated temperatures and high humid environments. The iso and ortho resin systems may be formulated with a wide range of fillers; including cellulose, mineral, short and long glass fibers. The ultimate in electrical properties is obtained by using synthetic fiber fillers such as orlon, dacron, or nylon flock. However, these materials are expensive and have high mold shrinkage. These materials must be molded by the transfer method due to difficulties in flash removal.

Electrical properties such as high dielectric strength, very low power factor readings, low dielectric constant over a wide range of frequencies, and comparatively long period of arc resistance are outstanding characteristics.

Parts molded from diallyl phthalate compounds are extremely stable and have a very low after shrinkage, especially if glass fillers have been incorporated.

Molded ortho diallyl parts will withstand continuous heat environments of 150°C. while the iso types are satisfactory for at least 175°C.

Diallyls may be produced in a wide range of opaque colors which are stable under most conditions.

Diallyl phthalates' biggest drawback is their relatively high cost per lb. Price may vary, dependent on the filler, from 86¢ per lb. to $3.00 per lb. Gravity may vary from 1.67 to 1.78.

Typical applications are: relay switches, connectors, terminal strips, electronic parts, and many other military applications where zero defect programs and electrical reliability are of utmost importance.

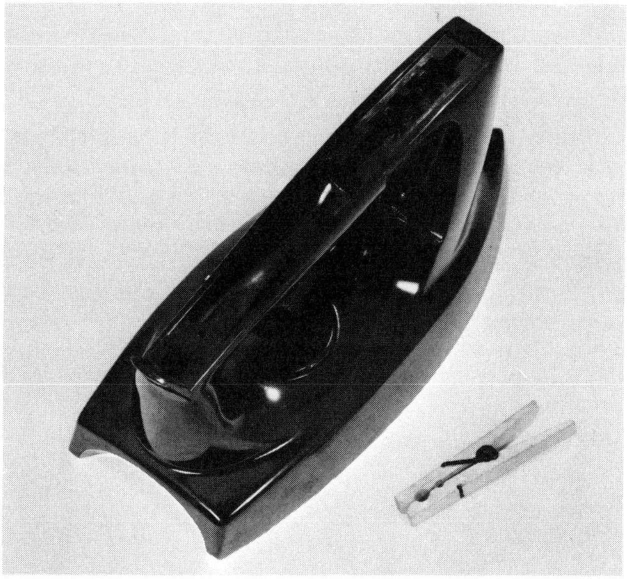

Figure 2: Steam Iron

Amino

Melamine molding compounds are produced by processing the melamine resin with various fillers such as wood flour, cellulose, and color pigments and release agents. The large volume materials are produced in granular form. When high strength is required glass fibers are used as fillers. However, the raw material becomes extremely bulky and somewhat hard to handle.

Melamine molding compounds tend to cure at room temperature, and consequently have a relatively short storage life. Molded parts also continue to cure after molding, and as a result have a rather large after shrinkage which may cause cracks to occur or parts to distort. Because of curing rate it may be somewhat critical to mold in parts having uneven wall thickness.

Parts molded from melamine compounds have a hard scratch resistant surface, practically no odor or taste, good electrical properties, fair arc track resistance, decorative appearance, and color fast characteristics.

The specific gravity range is from 1.45 to 2.00. The principal use of melamine compounds has been in the dinnerware field. They are widely used in electrical switch gear applications where arc track resistance is very important. In some cases the melamine phenolic compounds are used to produce electrical switch gear where arc track requirements are not so severe. Other applications are ignition systems (distributor cap), housings, and other decorative applications.

Urea molding compounds are generally produced using a urea formaldehyde resin with cellulose and wood flour fillers, lubricants, and coloring agents. They too cure at room temperature—consequently short shelf life. Molded parts continue to cure after molding, resulting in poor dimensional stability and extreme after shrinkage. The raw material is granular in form and can be molded on automatic equipment utilizing volumetric loading devices.

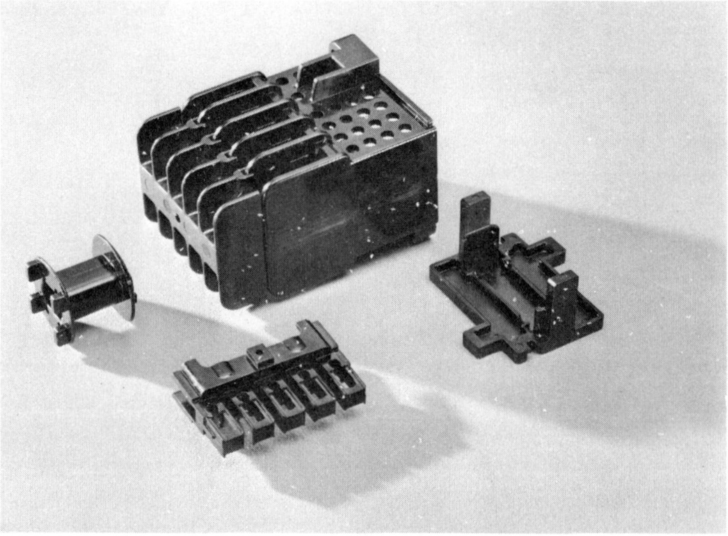

Figure 3: Motor Starter Switches

Parts molded from urea compounds have a hard surface, fair mechanical strength values, fair arc resistance, are resistant to solvents, oils, and greases; and are color fast if not subjected to direct sunlight.

The specific gravity of molded parts is 1.42–1.52, and raw material cost may vary from 23¢ to 40¢.

Typical applications are buttons, closures, electrical wiring devices (plugs, wall plates, outlet boxes, and wire nuts), electrical appliance housings, stove hardware, and other decorative parts.

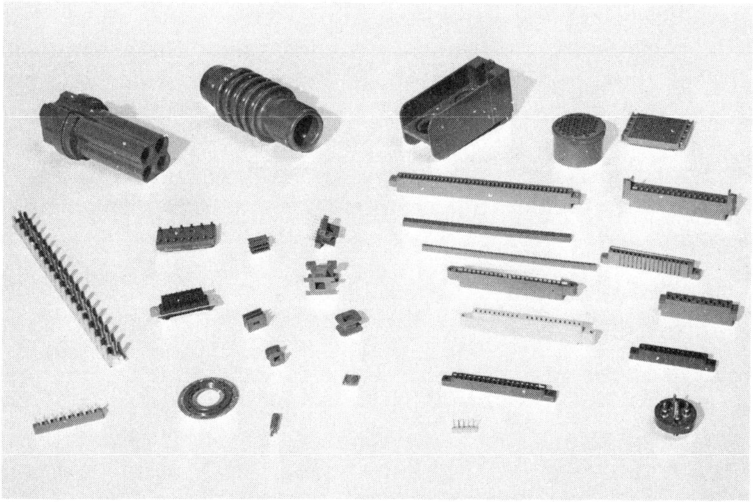

Figure 4: Electronic Components

Epoxy. Epoxy resins are basically the reaction product of bisphenol-A and epichlorhydrin. Molding compounds incorporate resins, fillers such as cellulose, wood flour, mineral, and glass. These molding compounds have very low viscosity ratings and may be molded under very low transfer pressures in the range of 50 - 150 p.s.i. Molded parts produced have comparatively low shrinkage, low water absorption, good electrical, and good chemical resistance.

Because of their very low viscosity under heat and pressure, epoxy compounds are used in encapsulated coils, electronic components, capacitors, resistors, and many other applications where delicate metal inserts or wires must be protected.

Compounds may be made in a wide range of colors which are relatively color fast.

Since the materials have a very soft plasticity rating, or inches of flow on spiral mold, the materials are all molded in special transfer molding presses and specially designed transfer molds. Because an acid catalyst and glass or mineral fillers are used to make the compound, these materials may have an adverse effect on the hardened steel mold with the result that there may be excessive runner and gate wear.

Specific gravity values may range from 1.49 to 2.0. Two main disadvantages may be the high cost of 80¢–$1.30 per lb. and the relatively short shelf life of some specific compounds.

Phenolic. Phenolic molding compounds, the oldest of the man-made materials, are produced having a wide range of properties, and are the most widely used of all thermoset molding compounds. Phenolic molded parts are hard, non-burning, rigid, strong, and dimensionally stable. They are resistant to low temperatures and 425°F. continuous use temperature, and two hour tests up to 550°F. In the case of missile applications they can stand extremely high temperatures for extremely short duration.

They may be produced to have excellent chemical resistance, superior electrical resistance, and impact strengths up to 17 ft. lbs. per inch of notch. Phenolic resins produced by combining phenol and formaldehyde in the presence of a specific catalyst may produce a one-stage or two stage resin. Many types of fillers are used such as wood flour, cotton flock, paper, mica, asbestos, various forms of glass, synthetic fiber (nylon), fabric filler, and many others. Dyes, lubricants, and plasticizers are the other ingredients. Molding compounds produced from a one-stage, two-stage, or combinations of both; specific types of filler; and color dye are so formulated to produce the desired characteristics of the end product.

Since it is possible to process materials having such a wide range of properties, phenolic materials are generally broken down into six general classifications: general purpose, non-bleeding, heat resistant, impact, electrical, and special purpose (including chemical resistant compounds). Having materials available which fall into the above classifications, naturally the applications are numerous. Phenolic molding compounds are used to produce parts in the following market areas: appliance, automotive (ignition systems, transmission parts, and power brakes), closure, communication, electric switch and starter gear, home laundry and dishwasher, power tools, textile, wiring devices, and a host of allied products. Specific gravity range is from 1.23 to 1.87. Cost range is from 22¢ to 60¢ per lb. with specials at increased cost.

Polyesters. Polyester molding compounds are generally formulated to produce parts having either good electrical values or physical properties with excellent chemical resistance. These compounds are produced in granular or premix form. Most common fillers are mineral and glass of various length fibers. Specific gravities may range from 1.65 to 2.3. Costs may vary considerably due to processing methods, grade of resins, and types of fillers. Applications are found in market areas such as: automotive, switch gear, aircraft, defense, chemical (pipe flanges, elbows, tees, etc.), and appliance (air conditioners, humidifiers).

Silicones. Silicone molding compounds are specialty compounds and are used in limited quantities. They are made using a silicone resin, combined with either glass or mineral. Molded parts made from these compounds exhibit high heat stability (450° - 600°F.), low water absorption, unaffected by ultraviolet, good chemical resistance, and good weatherability. In general, they have excellent electrical properties, resistance to corona, are non-tracking and non-burning.

Parts may be found in some coil forms, heating coils, and marine switch gear.

Specific gravities may vary from 1.6 – 2.8. The cost of the material is extremely high and is dependent on the specific formulation.

Processing Techniques

Thermoset molding compounds, when contained within a hardened steel mold, require heat and pressure to be polymerized into a solid mass. Molds may be heated by steam, electricity, or hot oil to temperatures of 280° to 425°F, depending entirely on the type of material and method of molding. Molding pressures may vary from a low of 50 p.s.i. to 15,000 p.s.i. Epoxy materials will mold at 50 p.s.i.; whereas, phenolic fabric-filled material may require excessive pressures. Again, the method of molding dictates molding pressures.

The time required to cure the materials is dependent entirely on the method of molding, mold temperature, and material temperature when introduced into the mold cavity. There are now four basic methods of molding thermoset molding compounds: (1) Compression, (2) Transfer, (3) Injection, and (4) Extrusion—with the method most commonly used as rated. Very small quantities are processed by extrusion. All methods of molding may be done automatically or semi-automatically. In most cases, the injection method is practically all done automatically.

Figure 5: Automotive Brake Parts

It is now common practice to preheat material before it is introduced into the mold cavity or cavities. In the case of compression molding, ultra high frequency preheaters are used to preheat material to 180° –250°F. temperature. This may be done with granular material or material that has been processed into preforms. In the case of transfer molding, either preheated preforms are loaded into the transfer cylinder, or granular materials that have been preheated by use of a reciprocating screw contained within a variable heated barrel, are transferred into the transfer cylinder. The latter method is commonly referred to as the two-stage transfer method of molding. When injection molding thermosets, the material is preheated by use of a reciprocating screw within a variable heated barrel. However, in this process the screw is also used to transfer material into the cavities through an orifice, sprue, runner, and gate system.

The injection method of molding is rapidly becoming more popular. To date, only slightly modified materials have been used. However, new resin systems will be developed and used further to improve the molding cycles and reduce the cure time. Using elevated mold temperatures in the 400°F range and screw preheated material at 260°F., extremely fast cure and reduced overall cycles are possible. Cure cycles of 15 seconds with complete cycles of 20 seconds are now possible when molding fast cure phenolic materials in molds to produce parts having wall sections of 0.125 inch thickness. An identical part, having thick cross sections of 0.250 inch or over, may be cured in less time when using a thermoset material than the time required to cool a thermoplastic material. Reduced processing time naturally reduces both piece part cost and capital expenditure costs of molding equipment and molds.

The thermoset industry is regaining applications lost to thermoplastic materials (an example would be dishwasher pumps). At the same time, they are making inroads on metal applications where metal castings or diecast metals have been used for many years. These are indeed exciting and interesting days, experimenting and participating in new developments, in what was called a dying thermoset industry.

Literature Cited

(1) Jacolow, Melvin F., Managing Editor, "Modern Plastics Encyclopedia," McGraw-Hill, New York, 1969.
(2) Society of the Plastics Industry, "Plastics Engineering Handbook," Reinhold, New York.
(3) Swanson, Robert S., "Plastics Technology," McKnight & McKnight Pub. Co.
(4) Volk, M.C., etal. "Electrical Encapsulation," Reinhold, New York.

RECEIVED August 4, 1969

4

Thermoplastic-Thermoset Engineering Materials

GERT F. BAUMANN

Mobay Chemical Company, Pittsburgh, Pa. 15205

This group of engineering plastics shows some of the typical characteristics of both the thermoplastics and the thermosets. At normal use temperatures, they exhibit a cross-linked network structure relating them to the thermosets yet they can be easily processed by standard or slightly modified thermoplastic forming methods. The network structure of the thermoplastic-thermosets differs in its nature from typical thermosets as shown by three examples (1) the ionomers, (2) the styrene-butadiene elastomers, and (3) the polyurethanes. Physical and mechanical properties of the three groups of thermoplastic-thermosets are discussed pointing out their special advantages as compared with competitive materials. Although the thermoplastic-thermosets are slightly more expensive than the largest volume thermoplastics or thermosets on a per pound basis, it is shown that the thermoplastic-thermosets are often favored because of their unique physical properties combined with cost savings in processing.

The thermoplastic-thermosets, as their name indicates, show some of the typical characteristics of both the thermoplastics and the thermosets, which were discussed in the previous papers. The group of polymers falling into this intermediate category is still small but steadily growing. The thermoplastic-thermosets were developed in an effort to combine several of the advantages of both the themoplastics and thermosets in one polymer. The properties of the thermoplastics are governed by their linear macromolecular structure, while the thermosets are characterized by a network structure of varying degrees of cross-link density. Therefore, the former can be melted at elevated temperatures and readily processed in the molten state by such techniques as extrusion or injection molding. Upon cooling, the thermoplastics return to their glass-like state. However, even in this glassy state, all thermoplastics, not withstanding their widely different physical properties, exhibit some plastic flow if an external stress is applied.

On the other hand, thermosets will not melt, but rather decompose at elevated temperatures, because their network structure is made up of irreversible primary bonds. At room temperature the thermosets show no or only insignificant cold flow under moderate to high stresses. The elastometric ploymers, which also belong to the class of thermosets, of course, differ considerably from the rigid thermosets in their physical properties. The former are characterized by their reversible elastic behavior as exemplified by high elongations and low moduli while the latter possess low elongations but very high compressive strengths.

We then see that a cross-linked network structure is a prerequisite for the absence of plastic flow and for high elasticities, while the linear polymers will flow under applied stresses, especially at elevated temperatures. From a processing standpoint, the thermoplastics are favored because they lend themselves to modern high-speed processing methods, but many of the thermosets' physical properties are more desirous than those of the thermoplastics. These considerations led to the development of thermoplastic-thermosets which possess a network structure throughout the most important use-temperature range, thus exhibiting physical properties similar to the thermosets. At elevated temperatures, these network structures can dissociate reversibly, resulting in linear macromolecules with their typical thermoplasticity. This permits processing by standard or only slightly modified thermoplastics' techniques. When the thermoplastic-thermosets are cooled, their network structure is reformed and the polymer behaves like a thermoset material again. The previously-described behavior of the three classes of plastics is diagrammatically summarized in Figure 1 which shows modulus-temperature curves for some typical polymers. Because of the extremely large variety of monomers and their possible combinations, Figure 1 can only be a typical presentation of the extremes, realizing that the polymers form a continuous spectrum with no sharp delineations.

We have chosen three groups of thermoplastic-thermosets to illustrate the principle of this class and to compare their processing, properties and applications with some thermoplastics and thermosets. There are a number of other polymers which might also belong in this group, at least in certain respects, but it is felt that the three groups selected are most typical and satisfy most of the premises set forth above. These groups are:

1. The Ionomers
2. The Styrene-Butadiene Elastomers
3. Certain Polyurethanes

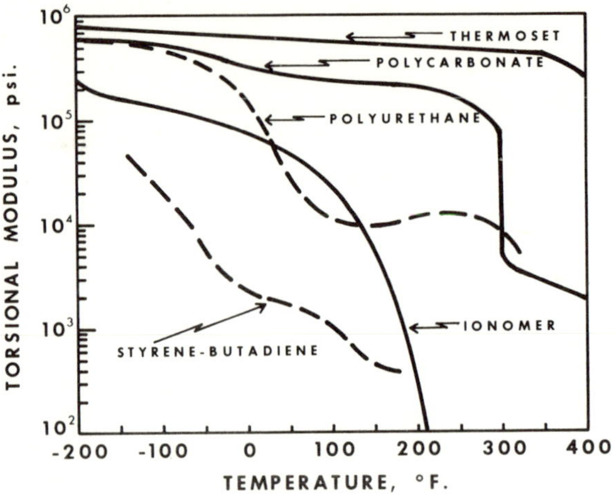

Figure 1. Modulus-temperature curves for various plastics

The chemical nature of these three groups differs vastly. Furthermore, their network structures are based on three different principles which is an indication of the large possibilities open to investigators for future thermoplastic-thermosets. The polymer backbone of the ionomers consists primarily of ethylene and a vinyl co-monomer, such as methacrylic acid, thus exhibiting pendant carboxyl groups. These linear chains are "cross-linked" by ionic, inter-molecular forces through incorporation of metallic cations from Groups I or II of the periodic table. Thus, the network structure is formed by electrostatic forces similar to those in inorganic crystals rather than covalent bonds as found in typical thermosets. The cross-link density can be varied in these systems by copolymerization of various amounts of vinyl monomer with the ethylene and by varying the type and amount of metallic cations in the polymer. As with all polymers, the physical properties are also influenced to some degree by the molecular weight and molecular weight distribution. A typical structure for ionomers is illustrated in Figure 2.

The network of the Styrene-Butadiene Elastomers is based on a second principle which, again, differs from the covalent cross-links of typical theromsets. The three-dimensional structure in this case is formed by physical cross-links which soften or "melt" at processing temperatures and reform upon cooling. The surprising stability of the physical cross-links is attributed to the specific conditions during the copolymerization which leads to a polymer with rigid polystyrene segments at the chain-ends connected by flexible, elastometric polybutadiene segments. The rigid chain ends form aggregates with the properties of pure polystyrene, thus stiffening the polybutadiene and acting like covalent crosslinks in vulcanizing elastomers, while the polybutadiene sections of the chain impart the flexibility, high extensibility and good low temperature properties.

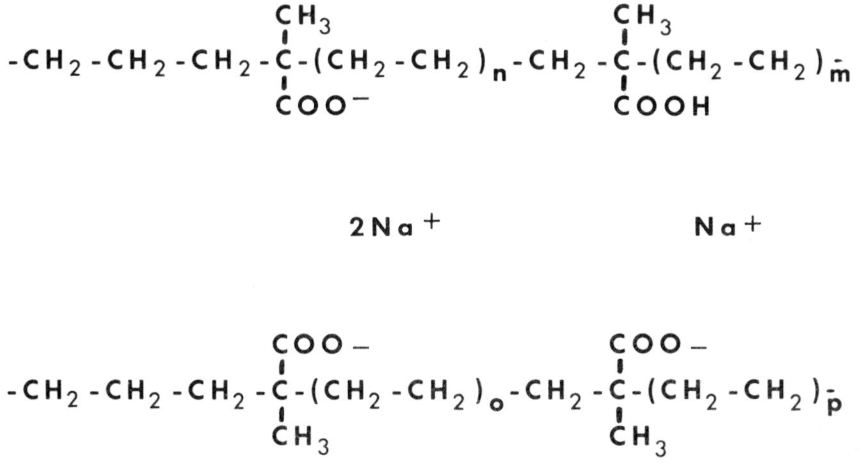

Figure 2. Ionomer

Although similar effects have been observed in highly crystalline thermoplastics, where the crystals act as "cross-links," the pseudo-thermoset behavior is more pronounced in the case of the styrene-butadiene elastomers. Furthermore, the formation of a crystalline structure is dependent on the processing and cooling conditions, while the polystyrene domains in the styrene-butadiene elastomers and thus the cross-links harden more nearly independent of the processing conditions. Figure 3 schematically shows the structure of the entangled network structure of the styrene-butadiene elastomers.

The Polyurethanes, the third group of polymers which will be discussed in this paper, represent a much broader scope than either the ionomers or the styrene- butadiene elastomers. Because of the large number of possible monomers which can be used for the preparation of polyurethanes and their respective combinations, the

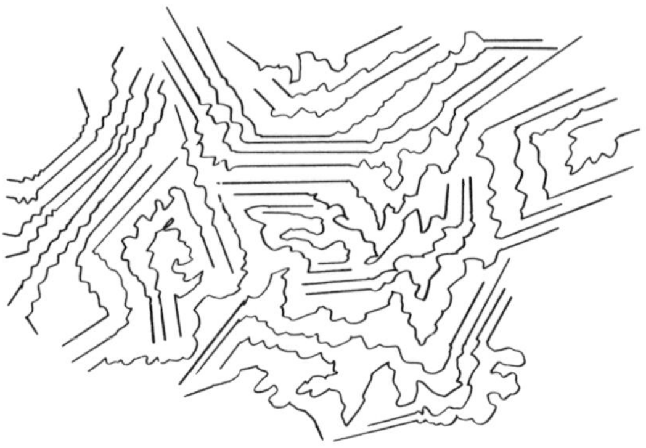

Figure 3. Styrene-butadiene elastomers

polyurethanes span the full spectrum of polymers from thermoplastics to highly cross-linked thermosets. The truly thermoplastic members are mostly of lower molecular weight and, of course, derived from linear starting materials while the thermosets are made from monomers with functionalities of three or greater. The thermoset polyurethanes owe their thermoplasticity to a combination of intermolecular hydrogen bonding and the formation of certain covalent bonds which can be reversibly dissociated at elevated temperatures. Figure 4 shows some possibilities for hydrogen bonding between groups typically found in urethanes. This is not to indicate that all these possibilities would be present in any single polymer. Figure 5 shows the two most common covalent bonds, the allophanate and biuret group, which can reversibly dissociate at processing temperatures and reform at lower temperatures. Under certain circumstances the uretdione group may also split thermally into isocyanate and then recombine with two urethane groups to form two allophanates.

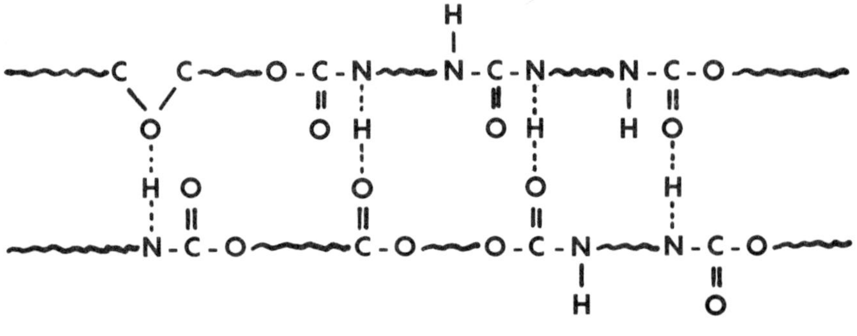

Figure 4. Possible hydrogen bonding in polyurethanes

So far, the definition of the thermoplastic-thermoset engineering plastics was based on their molecular structure and it was explained why these materials show similarities to both the thermoplastics and the thermosets. However, it is also possible to define the thermoplastic-thermosets strictly from a processing-property standpoint which results in a theoretically less well founded classification but a highly practical one. By this definition the thermoplastic-thermosets are those polymers which can be processed in short cycles, like thermoplastics, but also at low temperatures and relatively low pressures. Upon completion of the forming cycle, the polymers exhibit the typical properties of thermosets. The thermoplastic-thermosets defined this way differ from the ionomers and styrene-butadiene elastomers in that the former are processed as low molecular weight, low viscosity materials which are highly reactive and polymerize during the forming cycle to become thermosets. This process is not reversible and the final polymers show only very limited plastic flow. Their raw materials differ from those of most thermosets in the rate of polymerization and the low temperatures and pressures required for their molding cycle. The term "Reaction-Injection-Molding" has been coined for polyurethane engineering plastics in this group. As this name indicates, they are injected into a mold very similar to injection molding but, because of their low viscosity as compared to thermoplastics, they require very little pressure and can be processed at, or slightly above, room temperature.

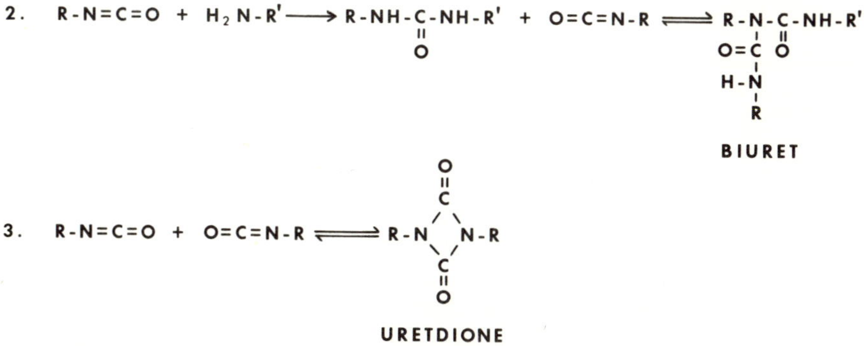

Figure 5. Labile bonds in polyurethanes

Processing

As was pointed out several times, the thermoplastic-thermosets can be processed by generally the same methods as the thermoplastics—*e.g.*, by extrusion, injection molding, or blow molding. Sheets of ionomers and polyurethanes can be thermoformed. The three groups under consideration in this paper can be handled on conventional thermoplastics machinery without major modifications. Some of the more pertinent conditions are briefly summarized in Table I. More specific information is available in the literature or from the manufacturers. It is significant to point out that the ionomers do not dissolve completely in any known organic solvent at room temperature. They are, therefore, not suited for solvent casting techniques. The same is true for certain polyurethane elastoplastics, while the styrene-butadiene elastomers can be processed from solution. Some special grades of polyurethane elastoplastics, mostly of highly linear structure, are available for solution casting, but these will not be discussed here.

Table I.
Processing Properties of Selected Thermoplastic-Thermosets

	Ionomer[a]	*Polyurethane*[b] *Elastoplastic*	*Styrene*[c] *Butadiene Elastomer*
Injection Molding			
Cylinder Temp., °F.			
Rear	350	370-400	300-350
Front	450	375-410	350-425
Nozzle Temp., °F.	400	385-410	425
Injection Pres., p.s.i.	10-20,000	10-15,000	15-30,000
Mold Temp., °F.	(40) 50-70	50-120	---
Extrusion			
Screw L/D Ratio	24:1	24:1	
Screw Comp. Ratio	3.4:1	2.5-3:1	
Screw Channel Depth	Very Shallow	shallow; long metering zone	
Melt Temp., °F.	380-480	390-440	

(a) Surlyn A, E. I. Du Pont de Nemours
(b) Texin Mobay Chemical Co.; Estane, B. F. Goodrich
(c) Kraton, Shell Chemical Co.

The "Reaction-Injection-Molding" process is schematically shown in Figure 6. In this case, the two monomeric or low molecular weight reactants, a polyisocyanate and a polyol, are accurately metered through separate pumps and mixed in a very small mix-

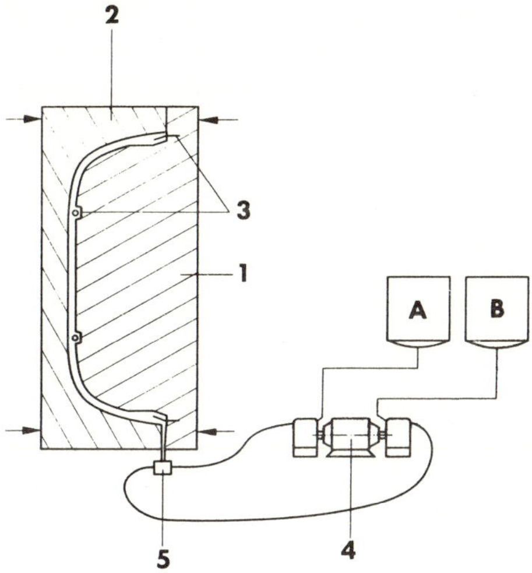

Figure 6. Schematic of the "reaction-injection-molding" process
A and B: Raw material tanks
(1) and (2) Mold; (3) Reinforcements; (4) Raw material metering unit; and (5) Raw material mixing unit

head by counter current injection under high pressure, and finally discharged into the mold cavity at ambient or slightly elevated temperature. Since both components are low viscosity liquids, no pressure is required to fill the mold. If the reactants contain an inert blowing agent to reduce the overall density of the finished article, a slight pressure will develop in the mold owing to the gas expansion under the exothermic conditions. This pressure, however, is only in the order of a few pounds per square inch, which does not necessitate the high clamping pressures encountered in injection molding. This allows the production of heavy parts with very large surface areas. Since machines with pumping rates up to 400 lbs./min. are commercially available, it is possible to make parts weighing about 75 lbs. with a single machine. The "Reaction-Injection-Molding" principle also lends itself, at least under certain conditions, to the use of two dispensing machines working in unison with a corresponding increase in output.

Polymer Properties

It is not the intention to give a complete summary of all the physical properties for the large number of specific thermoplastic-thermoset products available from various suppliers. These properties are aptly described in the literature. It is, rather, attempted here to point out specific properties noteworthy or unique for the ionomers, styrene-butadiene elastomers, polyurethane elastoplastics and the polyurethane structural foams which are processed by "reaction-injection-molding." We will also present a comparison between these properties and those of some competitive materials.

The styrene-butadiene elastomers exhibit the typical properties of vulcanized elastomers, but have outstanding low temperature properties and high coefficients of friction at a given hardness. Commercial products are available in various Shore A hardnesses with an attendant difference in modulus. At room temperature they also resemble some plasticized vinyl chlorides except that the elastomers are more resilient and show a much higher elongation and lower elongation set at break as evidence of their crosslinked network structure. Styrene-butadiene elastomers have a lower torsional modulus than either natural rubber or SBR and considerably lower than plasticized PVC at −20° to −100°F., thus, making them valuable materials for applications requiring flexibility at sub-zero temperatures. Table II summarizes the more important mechanical properties of the elastomers and gives a comparison with SBR and plasticized PVC.

Table II.
Property Comparison: Styrene-Butadiene-Elastomers vs. SBR and PVC

	Styrene-Butadiene Elastomers [b]	*SBR*	*Plasticized PVC*
Hardness, Shore A	40 - 85	45	40 - 70
Tensile Strength, p.s.i.	900 - 3000	2100	1200 - 2100
Elongation, %	500 - 1300	800	250 - 350
Modulus at 300% p.s.i.	200 - 1100	300	-
Elongation Set, %	15 - 80	30	120 - 150
Rebound, % (Falling Ball)	55 - 65	50	$<$10
Flexibility Temp.[a] °F.	−50 - −90	−25	30 - 40

(a) Temperature at which the torsional modulus is 10^4p.s.i.
(b) "Kraton" Elastomers, Shell Chemical Co.

The ionomers are outstanding for their optical properties of transparency and high gloss, their low temperature toughness and high notched impact strength, good abrasion resistance, resistance to organic solvents and to solvent induced stress cracking, and their high melt strength at processing temperatures. There is, however, a rapid decrease in modulus with increasing temperature between 100° and 180°F., limiting applications to the temperature range below 130°F. The room temperature modulus of the ionomers is equal to that of medium density polyethylene and certain plasticized PVC formulations, but approximately a factor of ten lower than rigid thermoplastics, such as, acrylics, polystyrene or polycarbonate. A summary of the properties is given in Table III which also shows the properties for medium density polyethylene, plasticized PVC and polycarbonate. Polycarbonate was included in this comparison because it also exhibits good optical properties combined with high toughness, even at low temperatures. Polycarbonate, however, has a higher modulus and its continuous use temperature is 200°F.

Table III.
Property Comparison: Ionomers vs. PE, PVC and Polycarbonate

	Ionomer[a]	*PE Med. Dens.*	*Plastic'd. PVC*	*Poly-*[b] *carbonate*
Tensile Strength, p.s.i.	3500-5500	1200-3500	3500-3700	8000-10,000
Tensile Modulus, 10^5 p.s.i.	0.5	0.25-0.55	1.1-1.2	3.5
Elongation, %	350-450	50-600	120	100-130
Izod Impact Str., ft. lbs./in. of notch	6-15 (½" X ½")	0.5- 16	Depends on type and amt. of plasticizer	14-17.5 (½" X 1/8")
Specific Gravity	0.93-0.96	0.93-0.94	1.32	1.2
Heat deflection Temp. °F., (66 p.s.i.)	100	120-165	-----	270-290
Brittleness Temp., °F.	−160	−100	+50	−150
Transmittance, %	75-85	10-25	Opaque	82-90
Organic Solvent Res.	Very Resistant at 75°F.	Resistant below 140°F.	Resistant to alcohols & aliph. hydrocarbons; swells in esters, ketones and aromatic hydrocarbons	Limited Resis.

(a) Surlyn A, E. I. Du Pont de Nemours
(a) Bakelite Ionomer, Union Carbide Corp.
(b) Merlon, Mobay Chemical Company.

As mentioned before, the thermoplastic-thermoset polyurethanes are divided into two groups for the purposes of this paper, those which have truly thermoplastic-thermoset properties, the "Elastoplastics", and those which can be processed by the "Reaction-Injection-Molding" process but are thermoset polymers in terms of their mechanical properties. The greatest commercial interest in the "Reaction-Injection-Moldable" polyurethanes at the present time is in the area of structural foams, typically in the density range from 8 to 50 lbs./ft.^{3} and the discussions here will be limited to this group, although it is entirely possible to produce the corresponding solid polymers without the addition of blowing agents using the same molding technique.

The elastoplastic polyurethanes have gained considerable commercial interest because they are distinguished by their extreme toughness, high compressive strength, low temperature flexibility, good abrasion resistance and inertness to hydrocarbon fuels and oils. The above property advantages can be obtained over a large range of hardnesses which make them unique elastomeric engineering plastics. The mechanical properties are summarized in Table IV which shows data for the two extremes with other products of intermediate properties also available. Table V shows the excellent abrasion resistance of elastoplastic polyurethanes as compared with a variety of rubbers and thermoplastics with similar properties.

Table IV.
Properties of Elastoplastic Polyurethanes[a]

	Shore Hardness Range	
	75 - 85 A	*50 - 60 D*
Ultimate Tensile Strength, p.s.i.	5000-7000	5500-7000
Tensile Modulus, p.s.i. — -100%	600-800	1800-3000
-300%	1000-1400	3200-4200
Ultimate Elongation, %	500-600	250-500
Elongation Set, %	20-30	40-80
Split Tear Strength, p.l.i.	250-400	200-750
NBS Abrasion, % of Natural Rubber Standard	150-200	

[a]The data include polyester and polyether-based polyurethanes:

Texin, Mobay Chemical Company
Estane, B. F. Goodrich Chemical Company
Roylar, Uniroyal Chemical Company

It is impossible to list representative properties for the "Reaction-Injection-Molded" polyurethane structural foams, because their properties can be varied over a wide range depending on the specific application. The foams can be very tough, resilient, blown elastomers or highly rigid structural foams. Furthermore, very recent developments have led to foamed structures with integral skins on the surfaces. These foams, in fact, are sandwich structures where the solid facings are chemically identical to the foam between the skins, thus avoiding corrosion, stress concentration at the skins, or delamination. In Table VI we have summarized some properties for a semi-flexible, high density foam, such as might be used for automobile exterior shock absorbing items. Table VI also shows data for high density rigid foam as it is employed in structural applications. Lower density foams are not listed because they were not considered engineering materials in the context of this symposium. As one would expect, the strength properties of the foams are highly dependent on their density so that only ranges can be given. In actual use, the foams are specifically designed to meet the properties required for the application. This gives design engineers a further degree of freedom in selecting the best material for the end use.

Table V.
Abrasion Resistance[a] of Various Polymers

Polymer	*Weight Loss, 10^{-3} grams*
Polyurethane Elastoplastic	0.5 - 3.5
Ionomer	12
Nylon 6,6	58
Impact PVC	89
Nylon 6	104
Natural Rubber (Tread Formulation)	146
Styrene-Butadiene Elastomers	125 - 165
Styrene-Butadiene Rubber (Tread Form.)	181
High Impact Polystyrene	545

[a]ASTM C-501; Taber Abrasion CS-17 Wheel, 1000 grams load. 5000 Revolutions.

The physical properties for Reaction-Injection-Molded urethane foam cannot readily be determined because they are, as all sandwich structures, highly dependent on the particular shape of the article and the ratio of skin to foam or, in other words, the density distribution through the part. Typical densities for these new structural foams are 30 – 45 lbs./ft.^{3} with average moduli of elasticity of 120,000 – 250,000 p.s.i. The compression strengths are higher than those shown in Table VI for rigid foams without skins.

The economics and applications for thermoplastic-thermoset engineering materials cannot be treated as a whole since the aforementioned three classes of materials differ so widely in their properties from highly elastic, rubbery polymers to highly rigid materials. The styrene-butadiene elastomers are used for shoe soles and heels which can be injection molded onto the shoe uppers. They have also found use in several sporting goods, such as, swim fins and masks. A variety of smaller industrial parts made from styrene-butadiene elastomers includes gaskets, pipe seals, grommets and bumpers. Sheet and simple shapes are produced by extrusion.

Table VI.
Typical Properties of Polyurethane Engineering Foams

	Semi-Flexible Foam With Integral Skin	*Rigid Foam*
Density, lbs./ft.^{3}	40 - 50	10 - 40
Tensile Strength, p.s.i.	350 - 850	250 - 2500
Elongation, %	150- 250	5 - 20
Compression Load Deflection at 50% deflection, p.s.i.	500 - 1500	-
Compression Strength at Yield, p.s.i.		200- 5000
Hardness Shore A,		
Skin Only	80	-
Foam Only	50	-
Impact Strength, ft lbs./in.	-	0.2 - 0.5

Special grades, meeting certain Food & Drug Administration regulations, permit the use of these elastomers for pharmaceutical items or those coming in contact with food. Although the cost of styrene-butadiene elastomers, 25 – 37¢ lb. with most grades at about 28¢ lb., is higher than that of many rubber stocks, the former are often more economical on a finished part basis because of low fabrication costs. On the other hand, thermoplastics which might be considered competitive generally show insufficient elasticity with attendant low elongations and high elongation sets. Plasticized poly (vinyl chloride) is lower in cost on a per pound basis but, owing to its higher density, about equal to the elastomers on a volume basis. Medium density polyethylene is a much as 50% cheaper.

The ionomers have found applications where the combination of clarity, toughness, elasticity, and solvent resistence are required, especially when these properties are demanded at low temperatures. Injection molded parts for refrigerators and other low temperature appliances have been made. Ionomers were chosen for automotive parts because of their resistance to petroleum products. The largest usages are seen in specialty packaging applications where the clear film is outstanding because it allows very deep draws and has excellent low temperature flexibility and toughness. The current price of 47¢/lb. or 1.62¢/in.[3] is considerably higher than certain low cost thermoplastics, such as polyethylene or PVC, but where the superior properties are required by the application this price difference may not be deciding. It must also be considered that the optical properties of most elastomeric materials are not comparable to the transmittance of the ionomers making the latter unique. A definite limitation in certain applications is the rapid loss in properties above 100°F.

The polyurethane elastoplastics are truly engineering plastics and have found many applications where extreme demands on the material are made. As some of the commercial applications will show, the claim that they are "elastomers which mold like plastics, work like rubber, and wear like metal" is not an exaggeration. The polyurethane elastoplastics are used for automotive front-end seals and have been a major factor in permanently greased ball joints. Many types of oil-resistant O-rings and gaskets are made from polyurethanes. Material handling rollers, especially where high abrasion resistance is required, have outlasted rubber rollers many times. Flexible couplings, parts on textile equipment, drive chains and belts, drive wheels for snowmobiles and various shoe soles and toplifts to mention only a few applications. A considerable amount of material in the Shore A 75 to 90 hardness range is used for extruded tubing or cable jacketing. In these applications it is primarily the ease of fabrication combined with the low temperature flexibility, good tear strength, and abrasion resistance which have led to the acceptance of the polyurethane. At the present time the elastoplastics are priced at $1.19 – $1.40/lb. Therefore, they are often used in those applications where design changes can reduce the amount of material required, where the high properties are necessary, or where costly maintenance can be reduced by switching to polyurethanes. In some applications the elastoplastics have replaced steel or other materials because of their better wear characteristics and elasticity, but in most cases, they compete against specialty rubbers and thermoplastics. They complement the well-known cast polyurethane elastomers permitting the fabricator to use high-speed forming techniques as opposed to casting.

The structural semi-flexible and rigid urethane foams, combining high strength with low density and low cost, have found wide interest and early acceptance in divergent markets although this development is still quite new. The semi-flexible foams are used primarily in the transportation industry. At medium densities, approximately 8–20 lbs./ft.^{3}, the foams are used either for decorative purposes or in recent years, as safety padding on the instrument panel, the A-pillars, on the steering column, in sun-visors, as arm rests, etc. The high density semi- flexible foams are employed in several front and rear bumpers on passenger cars. These bumpers will withstand low-speed impacts without permanent deformations owing to their elastic behavior. These bumpers also have allowed completely new styling concepts because they can be molded into more complex shapes than the conventional steel bumpers. The structural foams also can be used as shock absorbers, engine mounts and as elements in spring systems. The raw material costs between 25 and 60¢/lb. depending on the specific properties of the resulting foams. On a volume basis, they can readily compete with the lowest cost engineering materials due to their very low densities. Even at 40 lbs./ft.^{3} costs may be as low as 0.65¢/in.^{3}. Since it has become commercially possible to produce the semi-flexible foams with an integral skin, a skin consisting of the same urethane composition as the foam underneath, finished parts can be made in a single operation with an appealing, if desired, even grained surface. This, of course, eliminates several production steps thus further reducing the cost of the finished parts.

High density rigid urethane foams are a recent development based partly on technology gained from the well-known low-density rigid foams. While the latter have been used primarily for their superb insulating qualities, the former are used as structural members. The Reaction-Injection-Molding method allows high-speed production of various furniture components, some of which have very intricate designs simulating fine carved wood. If these parts are made with an integral, smooth surface, they can be readily painted without prior surface preparation. This concept is presently test marketed for institutional furniture. The rapid "injection-molding" process has given designers new possibilities for more modern and complex designs for chairs and settees.

The rigid urethane foam concept with integral skins has also found considerable interest in the automotive industry for exterior body parts, such as engine or trunk compartment lids, fenders and even full roofs. These parts can be reinforced with metal inserts where necessary or hinges and fasteners can be molded in for ready mounting.

The great interest that many industries have shown in these low cost high performance foams, which eventually will sell for well below 1¢/in.^{3}, will bring a very rapid growth to this latest segment of the urethane industry.

Literature Cited

(1) Bonotto, S., Purcell, C. L., *Mod. Plastics* **198,** 135 (1965).
(2) Eckerson, W. H., *Proc. 2nd SPI Intern. Cellular Plastics Conf., New York* (Nov. 1968).
(3) Gable, C. L., *Seminar on New Polymeric Materials,* Brooklyn Polytechnic Institute (June 7, 1968).
(4) Piechota, H., *Proc. 2nd SPI Intern. Cellular Plastics Conf.,* New York (Nov. 1968).
(5) Saunders, J. H., Frisch, K. C., "Polyurethanes: Chemistry and Technology," **Vol. 1,** Interscience, N. Y., 1962.
(6) Scheiner, L.L., *Plastics Technol.* p. 44 (March 1965).
(7) *Ibid.,* p.47 (April 1965).
(8) Shell Chemical Company, "Styrene—butadiene-Elastomers: Technical Bulletins. November 1968.
(9) Wirtz, H., *Proc. 2nd SPI Intern. Cellular Plastics Conf., New York* (Nov. 1968).

RECEIVED June 6, 1969

5

Applications Research

RICHARD S. DeBELL, BERNARD BAUM, and J. ROBERT FISHER

DeBell & Richardson, Inc., Hazardville, Conn. 06036

A qualitative discussion of the "spiral development" of a new polymer through successively more detailed rounds of synthesis, characterization, and evaluation for utility. As the understanding of the material deepens, its preferred areas for application are made apparent and can be explored specifically.

The arrangement of this book sets forth in logical sequence the assimilation of a new material into the commercial fabric of the plastics industry. Applications research represents the first tentative steps from the laboratory toward the market place, and provides the foundation for the market research and development to be described in later papers in this book. In its later phases, applications research may constitute the first step of market research, and this paper will defer to later speakers insofar as possible, on the techniques of sounding out the markets.

Applications research is directed toward finding uses for new polymers and plastics materials. One may well have reservations about embarking on this subject when the development of new polymers is reaching a state of diminishing returns both because they emerge less frequently and because they are progressively more expensive and chemically complex.

Applications research typically involves a spiral development which repeatedly reviews all the intelligence on plastics materials in progressively greater detail. The Air Force Materials Laboratory (2) program (Figure 1) provides for repeated cycling through materials modification, characterization, and evaluation. First, a preliminary characterization of the newly-created polymer will enable a comparison with the spectrum of existing materials, suggesting a range of uses. The preliminary characterization should be detailed enough to permit a first-round modification of the polymer. Having thus developed tentative targets for the material, it then becomes appropriate to make a more detailed characterization and ultimately to approach prospective customers to generate the management information necessary to define (a) whether to enter upon production of the new polymer, (b) in what volume to enter upon production, and (c) how much capital to invest in the pursuit of the new polymer. It is in the areas of preliminary characterization and in the accumulation of second-generation information for management decision that this paper will concentrate.

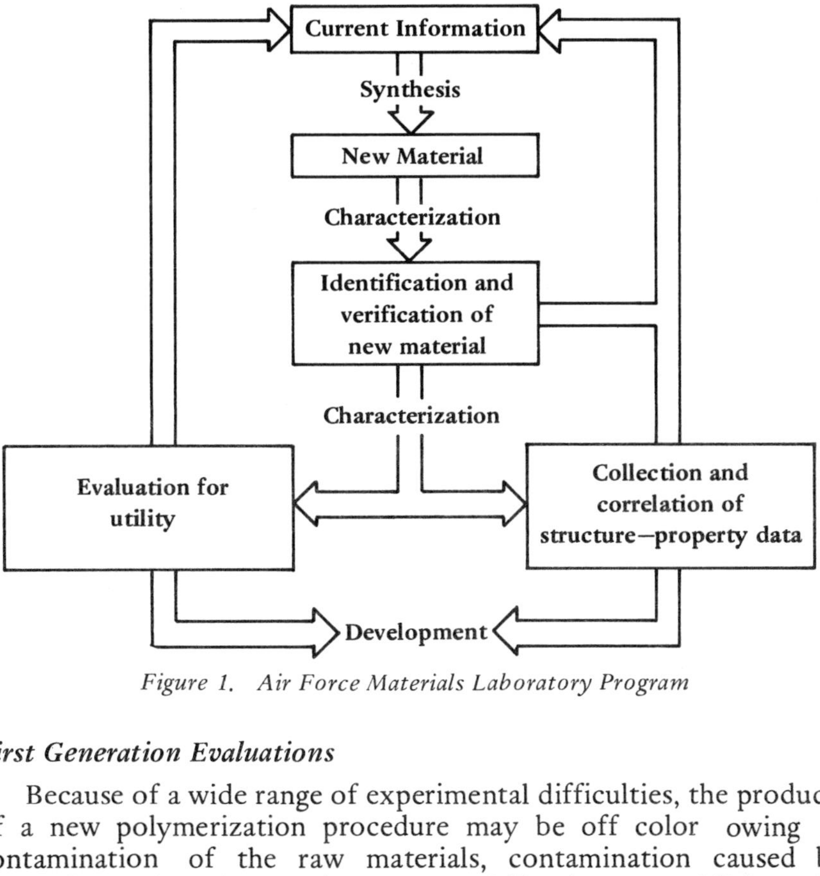

Figure 1. Air Force Materials Laboratory Program

First Generation Evaluations

Because of a wide range of experimental difficulties, the products of a new polymerization procedure may be off color owing to contamination of the raw materials, contamination caused by processing exigencies, or incorrect stabilization. In addition, the general polymer character may require adjustment by modification of the degree of polymerization or use of copolymerizing materials and by the addition of plasticizers, etc. In short, the researcher should be aware of the normal deficiencies of new laboratory-quantity materials and he should be prepared for and not discouraged by some rather glaring defects in the initial products.

It is important at an early stage to develop an understanding of the basic character of the product, in order to focus on properties of interest and thereby direct a refinement of the experimental polymer. Micro-evaluation techniques have been developed to a high degree over the years, and enable rather reliable characterization on gram quantities of new polymers. Unless pilot-scale equipment is cheap and readily available, and raw materials can be cheaply prepared, it will be important that the first evaluation be on the micro-scale rather than on pilot-scale lots since many experimental modifications may ensue and the cost and questionable repeatability of pilot scale quantities may conspire to put a serious financial "dent" in the program. Large numbers of recipes can often be prepared very cheaply in the lab; this is the reason for the development of the popular "bottle polymerizer," for example.

Several evaluative procedures have been devised, and they bear many points of similarity, differing mainly as appropriate to the special needs of each evaluator. Table I(2) lists properties studied in the Borg-Warner testing program of Chang and Goldberg. Table II(5) compares standard specimens with micro-scale counterparts; it is taken from a detailed article comparing values observed in standard and micro-tests on an impressive variety of materials, and generally confirming the validity of micro-tests.

The general melting characteristics can be established on a Fischer-Johns block or, in a more refined fashion, in a differential thermal analyzer. Should the DTA be available, it can be used on milligram quantities to define not only the melting behavior but also the temperature level and rate of thermal degradation.

A second characterization of major importance is a temperature-modulus curve. In this evaluation, a micro-tensile sample is clamped between the jaws of a universal tester, enclosed within a controlled environment, and gradually heated toward its melting point. At intervals during this heating, the tensile grips are separated and returned at a slow rate (0.2 inch/min.), and the spring rate observed. The resulting temperature-modulus curve can provide valuable indications of the tensile characteristics and service temperature of a material, and it has been observed that samples with a relatively constant modulus with increasing temperature may serve at higher temperatures than those in which the modulus decays with increasing temperature.

A third evaluation which can be applied to good effect to describe the processability of the polymer is a variable-load melt viscosity measurement. A precaution here is to conduct this test last, on the scraps left over from other physical testing, since the temperature in the rheometer may degrade the precious material irreversibly. A variable-load capillary rheometer simulates extrusion and may thereby provide the strand for evaluation of qualitative

Table I. Borg-Warner Micro-Testing Program, Material Characterization on a Sample of Less Than 20 Grams

Tests	*Material needed Grams*	*Tests*	*Material needed Grams*
Fundamental properties		Chemical resistance	
Intrinsic viscosity	0.1	Hydrolytic stability	
Specific gravity	0.3	Ammonium hydroxide	0.6
		Water	0.6
Mechanical properties		Sodium hydroxide	0.6
Yield strength	1.0	Sulfuric acid	0.6
Ultimate strength }			
Ultimate elongation }	0.2	Flammability (Informal) }	use left-over scraps from other tests
Young's modulus }			
Flexural strength }	0.6	Solvent resistance }	
Flexural modulus }		Hexane }	
Impact strength	1.5	Ethyl alcohol }	
Taber abrasion	4.1	Carbon tetrachloride }	
		Acetone }	
Thermal properties		Water }	
Glass transition temp.	0.4		
Softening temperature	0.2		
Heat distortion temp.	0.3	Solvent stress cracking	
		Acetone	0.2
Electrical properties		Hexane	0.2
Dielectric constant }			
Dissipation factor }	0.3	Aging properties	
Volume resistivity }		Ultraviolet aging	1.0
		Thermoxidative aging	1.0

properties such as smoothness of processability, the presence of volatiles, the strength of the melt, the draw-ability of a melt into fiber (for example) and a range of similar observations. Quantitative observations of the flow rate at a known pressure can be converted through the Poiseuille relationships (Table III) to afford first-round reference values of viscosity, shear stress, and shear rate for comparison with similar data available for example in Bernhardt, "Processing of Thermoplastic Materials" (1). Figure 2 (4) assumes a viscosity characteristic in this way from a single observation. It may be appropriate to emphasize that these are "apparent" values, since polymer flow characteristics are generally non-Newtonian. While melt viscosity can thus be analyzed in great detail, it is sufficient for the purposes of this paper to note that the capillary rheometer can provide good indications of processing limitations, on relatively small quantities of laboratory material.

Samples for a variety of physical, electrical, and chemical testing can be prepared by simple compression in a platen press, and thoughtful adaptation of standard test procedures allows valid

Table II. Specimen Dimensions

Property to be Measured	*ASTM Test*	*Standard Specimen Dimensions*		*Weight per Molding[a] (grams)*
		Molding	*Test Sample*	
Dielectric Constant & Dissipation Factor	D150-59T	2″ circular disk, 1/8″ thick or 4″ circular disk, 1/8″ or 1/4″ thick		10.24
Deflection Temp. Under Load (DTL)	D648-56	5″ x 1/2″ x 1/4″		10.24
Flexural Modulus and Flexural Strength	D790-61	5-1/2″ x 1/2″ x 1/4″ Same as for Flexural Modulus	5″ x 1/2″ x 1/4″ or 3″ x 1/2″ x 1/8″	
Tensile Elongation	D638-61T	8″ x 1-3/4″ x 1/8″	7.2″ x 3/4″ x 1/8″	28.68
Tensile Impact	D1822 61T		2-1/2″ x 3/8″ x 1/8″ or 1/16″ thick	
Tensile Modulus Tensile Strength	D638-61T	Same as for tensile elongation	Same as for tensile elongation	

[a] Assuming density of polymer equals 1.0
[b] ASTM Test D1708

measurements of dielectric strength, dielectric constant, gas permeability, creep resistance, ultraviolet resistance and chemical resistance on relatively small quantities of material. Note, importantly, that these are arranged in successive order of increasing likelihood of destructiveness to the plastics material, bearing in mind that stability is probably imperfect and some exposures are certainly destructive. The B. F. Goodrich micro-pilot plant(3) is an example of microscale polymer chemistry and evaluation on an organized, routinized basis.

At this point, a re-evaluation can be made of the projected cost for the monomers, the projected processing costs for preparation of the polymer, the range and limitations of polymer processability, the range of variability available in the polymer through chemical

(American Cyanamid Comparisons: A. E. Sherr)

Micro Specimen Dimensions		*Weight per Molding*[a]	
Molding	*Test Sample*	*(grams)*	*Remarks*
2-1/2" x 2-1/2" x 1/16" or 2-1/2" x 2-1/2" x 1/32"		6.40 3.20	Non-destructive test, can use sample for mechanical tests
Same as above	1-1/2" x 1/4" x 1/16" or 1-1/2" x 1/4" x 1/32"		
Same as above	Same as for DTL		Non-destructive test
Same as for flexural modulus			Destructive Test
Same as above	2-1/2" x 5/8" x 1/16" or 1/32" thick		ASTM D1708-59T
Same as above	2-1/2 x 3/8" x 1/16" or 1/32" thick		Destructive Test
Same as above	Same as for tensile elongation		Non-destructive test Destructive test

modifications such as the degree of polymerization or copolymerization or additive modification through plasticization, filling, etc., and the known characteristics of materials of similar properties or chemical analogy. All these points taken together enable a first decision on whether, and in what way, to proceed with the further development of the polymer type.

It is important at this point to give due consideration to the various ways in which a plastics material may be applied. 1. The material may be strengthened with additives to afford specialized properties such as hardness, abrasion resistance, lubricity or tensile strength. 2. The material may be extended either by fillers or by foaming (that is, filling with air) to overcome perceived economic disadvantage. 3. The material may be deposited in specific ways to take advantage of specific properties —*e.g.,* solution-coating of a highly impermeable but otherwise unprocessable polymer; alternatively; intractable materials may be made manageable by the addition of suitable lubricants or may require the development of modified processing techniques, for example in the case of ultra-high-molecular-weight polyethylene. 4. Physical modifications may be permitted, among which either uniaxial or biaxial orientation are principal examples. 5. It may find advantage in a peculiarity of processing — *e.g.* cast nylon and other materials suited to *in situ* polymerization may have an overwhelming advantage over materials more conventionally processed. In short, during this stage of the evaluation, considerable thought can be given to the means and the objectives of converting the material and making use of its unique properties or processing characteristics.

Table III. Viscosity Calculations Observations and Conversion Formulas

	Symbol	*Units*	*Comments*
Observed Values			
Flow Rate	Q	cu. in. / sec.	May require conversion from weight rate to volume rate thru specific gravity at test temperature
Applied Pressure	P	# / sq. in.	
Orifice Diameter	D	inches	Appears as fourth power in formulas. Measure carefully.
Orifice Length	L	inches	
Calculated Values[a]			
Apparent Viscosity	η	# sec. / sq. in.	Formula: $P\pi D^4 / 128QL$
Shear Stress	μ	# / sq. in.	Formula: $PD / 4L$
Shear Rate	$\dot{\gamma}$	$Sec.^{-1}$	Formula: $32Q / \pi D^3$

[a] Cylindrical orifice.

Second-Generation Evaluations

If the material has passed the first development hurdles of physical properties characterization, process selection, cost projection, etc., a further stage of refinement can be embarked upon. This normally would include the preparation of one or a few preferred polymers of the family, in reasonably stable form and under procedures designed to afford reasonably clean, nearly "commercial" product. Material will now become available in sufficient quantity and

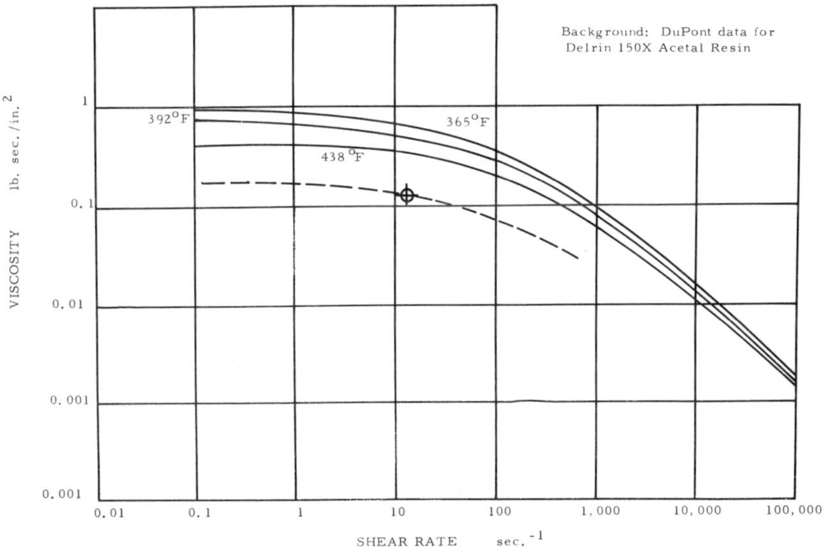

Figure 2. Single datum used for estimation of general flow behavior

purity that more detailed measurements of physical properties can be made both to insure that the initial promise is borne out and to extend understanding into other areas of possible advantage. Among the more important properties to be evaluated with the pre-production material are long-term properties as judged by accelerated aging or creep tests, sub- or super-normal temperature testing, and a detailed range of chemical exposures including particularly various chemical "ambients" to be expected in the areas of principal use – household or industrial. New tests—*e.g.*, the G. E. Limited Oxygen Index Test—are often introduced, and ease and refine the evaluative function. Strong emphasis can be placed on verifying the unique characteristics. As successive hurdles are passed and necessary corrections made, for example in stablization, additional experimentation can be undertaken with increasing confidence that it will provide useful background for the further stages of applications research.

Optimizing Field "Feedback"

By this time, a substantial body of information will be available to the applications researcher, and the influence of the polymer chemists and physicists begins to diminish. As he approaches potential users, the applications man should be well-armed with a first-hand familiarity with the physical and chemical properties and the processing characteristics of his new material. The attention gained in an interview with a prospective user is directly proportional to the amount of "home-work" done by the material developer. In addition to the paper-work, the applications researcher should have at hand a variety of specimens representing various converted forms of the material, as for example, typical extrusions or injection moldings, with which to stimulate the imagination of the prospective customer. The applications engineer should have a close familiarity with the applications and advantages and disadvantages of competitive materials, and should be prepared with striking visual displays or showmanlike routines for comparing physical and chemical properties with likely commercial competitors.

An item of major importance to the preparation of an applications campaign is a complete and easily understood checklist for the collection of all the information that will ultimately be needed by both technical and market-research personnel at his home office. This form should include space for observations on the total size of the market, the share enjoyed by the company interviewed, and the probable influence of selling price of the new material within that market. Only when the interviewer brings back information from which the technical obstacles, market size, and potential profitability can all be judged realistically on a preliminary basis—only then can a commercial development manager make the necessary decisions on the future for his material.

It is hopefully evident from the foregoing discussions that applications research follows in a progressive sequence of "spiral development" such that all aspects of the cost, properties, and applicability are reviewed in progressively more refined detail at each turn of the spiral. The process involves increasing rates of expenditure but allows the formulation of valid go:no-go decisions at several points along the way. Carefully organized, the expenditures can be optimized and the results of one phase factored neatly into the next subsequent phase as, for example, the final interviews will provide a sound basis for the market research work to follow. There is really nothing magical about the procedure, and all the steps are necessary to the orderly prosecution of a new material; depending on the impetus, the size of the steps can be varied and the expenditure rate suitably adjusted to fit particular needs. As the development progresses toward the ultimate commercialization, it becomes progressively more necessary to temper the optimism necessary at the laboratory stage with down-to-earth business judgement, since the ultimate judge, the customer, will be a discerning taskmaster worthy of your best efforts.

Literature Cited

(1) Bernhardt, E.C., "Processing of Thermoplastic Materials," Reinhold Publishing Company, New York, 1959.
(2) *Chem. & Eng. News.* p. 55 (Sept. 18, 1967).
(3) *Chem. & Eng. News.* p. 44 (Nov. 25, 1968).
(4) Richardson, P.N., *E. I. DuPont de Nemours & Co., Inc.* T. R No. 56 with additions (June 4, 1959).
(5) Sherr, A. E., American Cyanamid Co., "Preliminary Evaluation of Polymer Properties", S.P.E. Journal (*Soc. Plastics Engrs.)* 67 (January 1965).

RECEIVED June 6, 1969.

6

Marketing Research

JOHN F. AUCHTER

B. F. Goodrich Chemical Company, Cleveland, Ohio 44115

Properly conceived and executed marketing research, the "who, what, when, where, why, and how" of marketing, can be the key to the successful commercial development of engineering plastics. Its purpose, method of execution, and results will vary depending on the stage of development of the product. It is of greatest value when brought into the picture early in the life history of the product and continuously updated. By providing the right answers to the right questions, marketing research can guide commercial development from the initial exploratory research stages, through product development, application research, market development and into final commercial marketing.

Many of you are professional marketing research practitioners. Others are distinguished polymer chemists. All of us, at one time or another, produce, use, or evaluate marketing research results. The previous chapters have described the engineering plastics with which we are concerned here today. They have detailed their properties and general areas of application. I intend to cover how marketing research, given these parameters, can help guide the commercial development of engineering plastics. However, just as baseball teams started out spring training with a review of the fundamentals of fielding, hitting, and sliding, so perhaps a brief review of the fundamentals of marketing research may be in order.

Some of our self-styled social critics have claimed that "powerful new tools of modern management" (such as marketing research) have helped create a New Industrial State where new product success is virtually guaranteed.

I submit that there is nothing new about the function of marketing research—it was carried out by Bedouin traders at the dawn of history where it was literally a matter of life or death. It has been carried out in one form or another, as long as one man has been trading with another. (These same social critics also ignore the fact that, according to the American Management Association, some 90% of new products fail in their first year on the market—but that's another story.) My point is that marketing research must be done by someone. Commercial development, technical service, research and development, the field sales force—all have a role to play. Most chemical companies today have chosen to formalize this function and place the responsibility for its execution in a separate Marketing Research Department.

Purpose and Function of Marketing Research

What then, is this function of marketing research? There are many definitions, but the one I like best is—it is the collection, evaluation, and communication of whatever information is needed to assist management in making decisions in the marketing area. Another way of looking at it is to take a page from our friends in the journalism profession and say that it is the "who, what, when, where, why, and how" of marketing. Note the all important "how."

This is what puts life into what otherwise may be merely a sterile collection of data—it puts the "ting" into marketing research. Now let's see how this pertains specifically to the field of engineering plastics. Let's suppose we are starting out on a hypothetical marketing study.

Planning the Study

Any such study consists of three main parts—planning, executing, and reporting. The planning stage is obviously crucial. And perhaps the most important part of it is a clear statement of the problem and the desired objectives. The necessity of this is so obvious that it is often overlooked. However, in marketing research as in other disciplines, there is a point of diminishing returns, and the researcher must understand the intended use of his work to know when that point is reached. We must determine whether our hypothetical study is to cover just one material such as polycarbonate resins, a broad family of materials such as styrene based copolymers, or the entire range of engineering plastics. Or our study may not be oriented around materials at all; but its purpose may rather be to measure and evaluate unfilled needs. Should international factors be considered? This is vital in products like acetal resins where a large portion of production is exported; or for products like poly-4-methylpentene where the only commercial producer is located abroad. Each of these different studies will have different objectives and different plans of attack.

A problem common to all commercial development programs is: "At what point should marketing research enter the picture?" This really resolves back to the question of whether we develop a product first and then find markets for it or do we identify market needs and then develop a product to fill those needs. We all may favor the latter approach and extol the benefits of the marketing concept wherein we develop products to meet needs. But I think we must admit that much of our effort is based on finding markets for products our companies have the ability to make.

And this is not all bad. Who among us would have had the vision to foresee the use of engineering plastics to meet the need for the hula hoop or the stickless, greaseless frying pan? But development work was done on high density polyethylene and fluorocarbons because of a basic faith, a "gut feel," if you will, of the inherent value of the sum total of the properties of these engineering plastics. Nevertheless, the earlier in the commercial development program that marketing research is brought into the picture, the more valuable it can be.

Ideally, it is utilized before a research and development project is started rather than after. Let's assume, in our hypothetical case, that we are asked to do a marketing research study on a polymer about to enter the pilot plant stage. Several other companies are already commercially producing this product and enjoying rapidly growing sales. Our product, made by a different process, will be basically the same, but will have several distinct property and processing advantages. I believe this represents a situation with which we are all familiar. Such a situation lends itself to the dual approach I like to call "Bottom Up and Top Down." The distinction between these two approaches will become evident during the remainder of this discussion.

The "bottom up" approach is based on typical marketing research field work. The analysis and projections are derived by putting together all the individual bits of information to form the whole. It is oriented around actual and potential users.

In our illustrative case, identification of these users should present no problem. We have an established pattern of use to start with. Typical uses of engineering plastics will be located in such industries as automotive, aircraft, appliances, business machines, communication equipment, aerospace, and defense industries. Our sales department can help guide us to the right companies in these fields. We should include also so-called secondary accounts – our customer's customers.

If it is decided to conduct a field survey and if your company does not have background in this area, considerable time and money may be spent screening the hundreds of companies you may wish to interview. At this point, it may be helpful to utilize one of the many companies that specialize in conducting telephone screening surveys. The level of sophistication of the questioner can vary from telephone operator to professional engineer as desired. Like any other marketing research operation, this must be carefully planned and supervised to be successful. Properly handled, this can be an inexpensive way of generating information.

Our applications research should be able to provide important help—particularly regarding new uses taking advantage of our special properties. Other organizations such as trade associations, labor unions, governmental bodies, and trade journals may also be important in particular cases.

After we have selected the companies we may wish to contact, the next question is: "What type of person or job function should we contact?" The level of commercial development of our product can be a guide. If we are still in the test tube stage, it may be premature to have extended discussions with the purchasing department. This may vary greatly among potential customers depending upon company policy and the personality and interests of the particular purchasing agent. It is far better to let him tell you to come back when the product is developed than to run the risk of offending this all important link. Our sales department should be a good guide as to how to handle each case. Generally, research and development people are likely candidates for marketing research interviews. As commercial development of our product proceeds, the materials engineers and design engineers of the company become of increasing importance. I have found that in most cases a user's marketing department should be contracted regardless of the stage of development. The marketing realities of our customers' business will play an important and sometimes overlooked role in our product's future.

In our hypothetical case, we would probably want to see individuals representing research and development, marketing, and perhaps materials and design engineering. If the company has a marketing research department, your counterpart there is usually a good place to start for suggestions as to what individuals you should see. The number of calls can vary from three or four to several hundred, depending on the nature of the product and market and the degree of coverage desired.

Executing

Executing the interview should be the easiest part of the study—if the researcher has prepared himself properly. The questions to be asked will, of course, depend upon the nature of the study. The researcher should bear in mind that the interview should not be considered a one-way street, but rather an exchange of information. He should be prepared to offer non-confidential data on probable market size, property requirements, or material characteristics to ease the flow of information.

Above all else, he must remember that he is the representative of his company, in fact is the company in the eyes of the respondent. He is now, in part, a salesman selling not a product, but his company's image and reputation. He is in a position to do his company good or harm, depending on how he conducts himself. Marketing research, dealing as it does with vital questions and important customer relationships, is clearly not the place for a novice.

What I have been describing is the way a typical field marketing research study is organized—putting together many bits and pieces to form of total picture of the market. A complementary approach, what I call the "top down" approach, is to view the total market as a whole, then take a look at all competitive materials being used. Such an approach should consider not only engineering plastics, but also more traditional markets such as aluminum, zinc, magnesium, and even non-metallics such as glass and wood, where appropriate.

The degree of sophistication of such an analysis can vary all the way from a simple guesstimate of the relative merits of the competing materials to a statistical value analysis using multiple regressions. Such a procedure has been described in detail by S. T. Pender*(1)*. Even the so-called "panel of experts" approach may be useful. A word of caution here—just as in the field research, the market researcher must avoid becoming a mere collector of data and opinions.

Ideally, these two approaches, bottom up and top down, will mesh and reinforce each other. Suppose, however, they do not—what then? In most cases, I would choose the bottom up results. There is no substitute for deliberate, creative, carefully executed field marketing research. In the case of engineering plastics, however, there can be opportunities which all the field work in the world will not reveal—witness the example of the hula hoop. When these two approaches yield widely differing results, it is a clear sign that further work is necessary.

Reporting

Now we come to what is frequently the most crucial part of the entire study—communication of the findings. And this includes not only the presentation of data and analysis, but also conclusions, and most important, recommendations. At this point in his study, the market research analyst should be the best informed person in the company on the subject under investigation. As such, management wants, needs, and is entitled to his recommendations for further action. Without them, the report may be of purely academic interest.

As to manner of presentation, only two words of advice—an oral presentation is both a valuable supplement to the written report and an effective means of focusing management attention and securing action. And regarding the written report, brevity is not only the soul of wit; it may also be the key to a successful report.

Thus, marketing research, by getting the right answers to the right questions, should guide the commercial development of engineering plastics from the initial research stages, through product development, applications research, marketing development and into final commercial marketing.

Literature Cited

(1) Pender, S. T. Rept. to Chem. Marketing Res. Assoc., 1965.

RECEIVED June 6, 1969.

7

Process and Processing Development

WILLIAM J. COUGHLIN

UNIROYAL Chemical, Division of UNIROYAL, Inc., Naugatuck, Conn. 06770

Two characteristics of plastics are processability and cost. Processing is a key factor required to commercialize a product. A process must be developed to scale up economically from bench to pilot plant and eventually to a production process. Equally important is adaptation of new polymer to economical fabrication. Successful fabrication insures that laboratory properties are inherent in end product. A processability value must be included in initial objectives and product description. Test methods and procedures developed must correlate with commercial fabrication. These must be applicable to small amounts of available materials. Market development and application development expertise must be utilized throughout. Research and development objectives must be matched to market needs to insure the product's commercial value.

The successful commercialization of new, profitable technology depends on utilizing a broad range of talents and functional areas. In a large corporation it is almost impossible for one man to take a product from incubation to a full-scale profitable venture. Market research, market development, research and development, and engineering talents are all required before a potentially profitable new product is refined to the degree that it is ready for adoption by production and sales. However, many successful projects do have a single person as the champion or principal driving force. New technology starts with an idea and requires someone with vision, faith, and persistence to overcome all the obstacles that face a new venture.

The successful development of a new plastic material involves two steps: (1) an efficient process to manufacture plastic material and (2) development of efficient methods for fabricating or processing the material into useful forms. This second step may have to be divided into several steps, and it may be necessary to go back to the first step for modification in the plastic material before the material will meet the needs.

Finally, optimizing these processing steps is essential to a successful and profitable venture.

I would like to pursue the important steps, in sequence, that are necessary to develop a new plastic and to relate the importance of good processability and/or a good process to the ultimate success of a product. (*See* Figure 1).

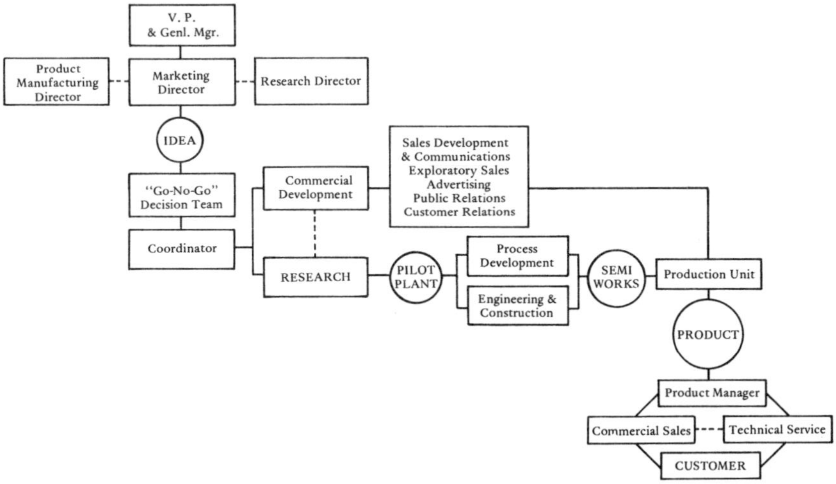

Figure 1. Anatomy of product development

Importance of Establishing and Setting Objectives

Clearly established and well-defined objectives are important guides to any individual regardless of the functional area or management level involved. An understanding of these objectives is absolutely essential to long-range planning, research and development, and market development activities. It is difficult enough under the best of circumstances successfully to develop profitable new products without wasting money and time developing a product that is undesirable for reasons that could have been predicted in advance by anyone who clearly understood company objectives.

There are many promising approaches to developing new technology, but one company cannot pursue them all. The most logical approach is to spend the time, money, and effort on those projects that are compatible with total company plans; in short, where top management will invest money.

The Function of Market Research

Market research is an important and effective link in the chain of events that starts with an idea and leads to an investment in research and development. Market research is required to ferret out market needs and to help describe the characteristics of the product that could fill this need. Many new plastics are expected to replace nonplastic materials in specific applications, and part of the input must include the developing of a suitable new fabricating process. Anyone involved in these activities knows that it is an oversimplification to expect market research people to go out into the market place and ask potential users what their future needs are going to be. A large majority just do not know, particularly when yours program requires an estimate of the profitability of a project before it is initiated. This requires not only an understanding of a need but also an appreciation of future pricing, volume, and costs.

As an example, one of our company objectives is to broaden our base in plastics through the introduction of engineering plastics. This seems to be compatible with our history of research, technical service, and technical sales. The term engineering plastic is a very broad, if not vague, term. As might be expected the results of some of our early market survey activities were quite general. Later our understanding of the needs of the automotive industry did help us to establish some research objectives. We believed that a polymer with the following characteristics would be useful: (*See* Figure 2) (1) "Dimensional Stability" at automotive touch-up paint oven temperatures which we expressed in terms of a 264 p.s.i. heat deflection temperature of 300°F., (2) "Good Processability" which we expressed in terms of a 450°F. plastic Mooney value, and (3) "Impact resistance or toughness" which can be measured by several conventional techniques.

CHARACTERISTICS	*WHY?*
High Temperature Resistance	To Withstand Heat of Automotive Touch-up Ovens
Good Processability	To Insure Optimum Properties and Economics
Impact Resistance or Toughness	To Perform under Automotive and Appliance Service Conditions

Figure 2. R & D project objectives (New Polymer)

Once we had clearly defined these objectives we were able to conduct a much more specific market survey. We did add one more important property, an estimated selling price. This program led to the development of a new plastic which will be discussed later.

A quick glance at Figure 1 will help to establish where we are in the product life cycle.

Research and Development – The Primary Source of New Technology

Once the initial planning has been completed the primary task of developing a new product or process can be undertaken. An ideal program will have clearly established project objectives along with the confidence that these goals are compatible with overall company objectives. The knowledge that there is a need for this sought-after technology increases the probability of financial success. After a product has been developed on the bench or in research, several very important factors must be resolved. As you all know, there may be several times during the early life of a new product when it is essential to develop an estimated factory cost before the potential profitability can be projected.

We try to introduce process development and engineering talents into product research programs as early as possible to help insure optimum low cost processes. We know that new equipment may have to be designed or evaluated as part of the program to develop efficient, low cost production techniques. Since these initial cost estimates are theoretical they are done primarily on paper (*See* Figure 3).

Fixed Capital Investment
Raw Material Cost
Chemistry, Processes, and Number of Operating Units Required
Equipment Selection
Utilities and Offsite Costs
Labor Content
Effluent and Waste Disposal Costs

Figure 3. Elements of a process cost

Later as a new product moves closer to commercialization the actual process development and engineering efforts must culminate in a reliable economic process that produces a product with the desired characteristics and quality. Anyone involved knows that this scale-up is expensive. It is time consuming to develop both a commercial process and the necessary supporting cost estimates. It is acknowledged that premature engineering or pilot plant activities can be costly if the polymer synthesis research work has not been finalized. This is where good judgement is required to insure the optimum timing of a project.

While this may seem basic, market potentials depend on the selling price of this new product. Profitability remains a function of the cost to produce this new product and the relationship of this cost to the selling price.

Skilled development personnel realize that processing is as important to the user as a consistent low cost process is to the producer. Early product specifications should include a test to measure the consistency and ease of flow or processability. To us the high temperature Mooney measurement represents an important research tool for predicting processability.

The ultimate profitability of a product depends on many costs, including the fabricating or processing costs as well as the producer's factory cost. Easy processing insures that the optimum properties can be developed at a minimum cost. This is a particular challenge to producers of high temperature-resistant engineering plastics because of their inherent resistance to deformation at high temperatures.

Perhaps this is an opportune time to cite a case history, which I was involved in, where processability was a key element of an important new product development program. Several years ago a few companies in the refrigeration industry were interested in improving quality and reducing costs and considered expanding the use of plastics to realize this objective. These manufacturers expressed an interest in ABS, one of the polymers Uniroyal produces. Initial evaluation proved that, while interesting, none of the grades of ABS available were satisfactory. One of the refrigerator manufacturers established, as a goal, Polymer X, a hypothetical plastic with the combination of properties they desired for refrigerator door liners and inner liners. These characteristics included impact strength over a wide temperature range, structural strength and rigidity, resistance to stress crazing in a broad range of environmental conditions, appearance, and processability. As might be expected, these desirable properties did not exist in commercially available plastics.

This technical challenge was compounded because it is very difficult and expensive to measure the processability characteristics of small experimental batches and to expect that these results will correlate with production experience. Performance and appearance of the final part is impossible to obtain without excellent and consistent processability which, in this case, includes both an extrusion and vacuum-forming operation.

Several times during this lengthy development project we felt that we had developed the proper combination of properties. Later we were disappointed by inconsistent or poor processability during plant trials. We found that even these very expensive trials, using 2-1/2 inch laboratory extruders, were not giving us the proper perspective. Extrusion conditions and the exposure times to processing temperatures did not correlate to production conditions. Considering the many polymer modifications required, it is very difficult and expensive to make enough experimental material for laboratory extrusion trials.

Finally, we developed testing procedures that included Brabender studies and the use of high temperature Mooney machines, and our program began to move. We were able to correlate with production sheet extrusion conditions which enabled us to make our numerous product development changes rapidly on a small scale.

A small "bubble test" was adopted that allowed us to evaluate performance and appearance after vacuum forming. This technique requires a small sheet sample which is heated under specified vacuum forming conditions and then blown into a bubble of predetermined size. Now our engineers were able to predict the appearance of the final vacuum formed part during a laboratory or plant sheet extrusion run.

On this particular project we were fortunate that the market demands were made known by some of the major refrigerator producers and the requirements were described by the characteristics of Polymer X. Nevertheless, a typically extensive research and development program was required followed by an intensive market development program after the product was developed. Our own process development was important throughout the product development program. It actually became most vital after our new ABS was developed and accepted. A low cost process was essential and the proper quality was assured only after special tests and instruments were developed. Maintaining a low volatile content was particularly critical.

Processing Development

An important part of any plastics research and development program is the time that should be spent on evaluating and developing new processing systems and equipment. The introduction of many new polymers and the dramatic growth of the plastics industry would not have been possible without the development and the commercializing of new fabricating concepts and equipment.

One of the outstanding characteristics of thermoplastics is that they can be fabricated by many different techniques. Development of the optimum processing system for a new plastic is often the key to the economics that lead to the use of plastics in a specific application. (*See* Tables I and II, Figures 4-9).

These Figures help to highlight the development of new equipment and its effect on the growth of plastics. Think of the importance to engineering plastics of the evolution that has taken place in the development of the injection molding machine! Currently there are 40 injection molding machines with 2000-5000 ton clamping pressures; sizes unheard of just a few years ago. Machines with 200-300 oz. plasticizing capacity are not unusual

Table I. Plastic Equipment And Concepts Some Outstanding Advances

Trends & Developments	*Significance*
Injection molding	
Screw machine	Permits the processing of engineering plastics
200-300 oz. Machine	Allows large parts to be molded Automotive parts, containers
2000-5000 ton Clamping Force	Now huge parts are possible Automotive body panels Refrigerator liners Material handling pallets
Extrusion process	
Tailored screw design	Optimum design is the key to any extrusion operation
Larger capacities Higher speeds – extruder	Faster film and sheet extrusion rates More economical wire coating Lower cost pipe
Blow molding	
Commercialization of a new concept	Plastic packaging and containers feasible
Availability of new equipment	Practical to consider large automotive and appliance parts
Vacuum forming	
Larger machines	Useful for low volume runs or where timing is critical
Greater productivity	Economics based on low tooling costs
Greater productivity	Interest in very large parts Campers, trailer bodies Automotive body parts Parts with thin walls

Table II. Plastic Equipment And Concepts Some Outstanding Advances

Trends & Developments	*Significance*
Cold forming	Opportunity to fabricate plastics the same way as metal parts New application opportunities Potential for Improved Economics
Roto Casting	Method for making large bulky parts with complex design Minimize assembly operations and costs
Rigid urethane foam lamination	Allows composites of rigid urethanes with rigid or flexible materials to be made on a continuous basis Metal, wood, ABS or PVC outer panels most promising
Metal plating process New concepts and chemicals	Combines plastics with chrome, nickel and other metals analogous to chrome plated zinc die cast parts Combines some of the best characteristics of plastics and metals.

today. Now manufacturers are considering shot sizes up to 100 lbs. (1,600 ozs), although 800 ozs. is probably the maximum size actually being designed. As important as these large machines are to large plastic parts, probably none of this application work would have been possible without the development of the screw injection molding machine.

Large plastics parts are important to the future growth of plastics, and it is interesting to speculate on the part to be played by blow-molded automotive parts and rotocasting large, bulky parts of complex design, a technique that minimizes costs and assembly operations.

Cold forming of plastics is a relatively new technique which introduces the opportunity to fabricate plastics similarly to metals.

Since engineering plastics are difficult to process under normal conditions and occasionally require special equipment, it is essential that commercial fabricating requirements be resolved early in the development program. This helps insure proper acceptance once the polymer is introduced to the market.

Proper paint and vacuum metalizing systems, metal plating procedures, welding, and adhesives are some of the other important considerations in the fabricating process that must be factored into a program before a new plastic can be fully accepted.

To summarize, let me quote from a comment I heard recently. The president of a large corporation expressed the thought of many

Figure 4. 4000 ton injection press

top executives when he stated "Successful research is essential not only for growth but for sheer survival." Uniroyal is one of the many companies that tries to meet this challenge by introducing a steady stream of new, profitable products to help support corporate growth objectives as well as generating process technology to insure competitive, low cost production.

Figure 5. 350 ounce injection molding

Figure 6. 76 inch diameter E. B. Blue Co. rotational casting machine

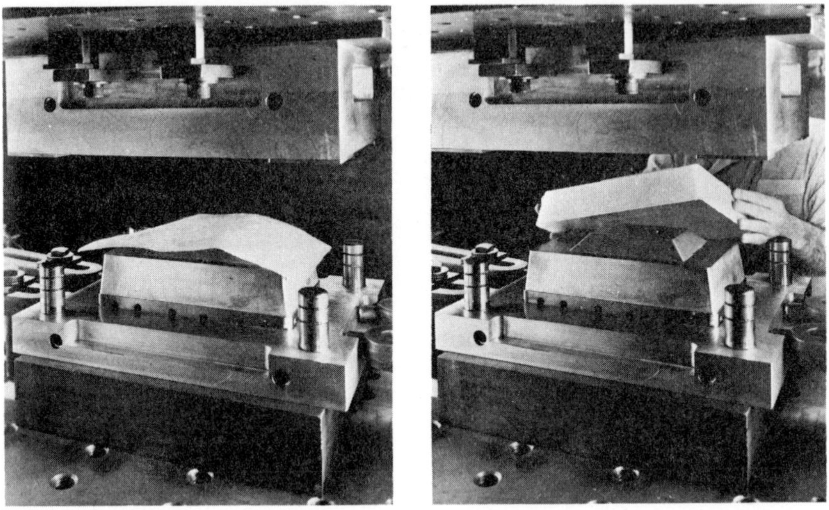

Figure 7. Cold forming

It is essential that research scientists and engineers understand the more specific objectives of an individual project, as well as the broad corporate objectives. They should not be unduly restricted by being told how to reach an objective, but they should clearly understand what these objectives are.

Another look at Figure 1 shows where we are in the life cycle of a new plastic and the next phase, Market Development.

Another Essential Aspect in Developing a New Polymer–Market Development

Some new products can be sold immediately, particularly if they are closely related to present product lines or are developed to meet competition. In the case of a new plastic you never know whether you have a good product until someone is willing to pay for it. A product can be developed, the market located and described, and a profitable future predicted, but this will all be an academic exercise unless the product is sold.

The market development function is essential because many plastics are so new and different that they must be introduced carefully: (1) it must be demonstrated that you have a product with performance and consistency, verified by commercial evaluations; (2) consistent processability must be demonstrated; (3) initial availability is probably limited by pilot plant or semi-works capacities; and (4) the end users must be sold on the performance of a product and specific applications developed.

Figure 8. Cold forming

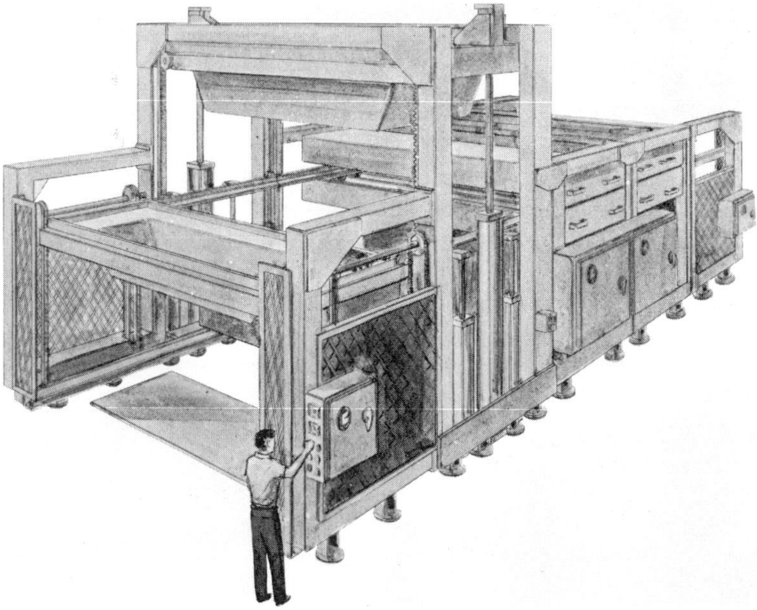

Figure 9. Large thermoforming equipment

Now to the specific case history I alluded to earlier. During 1968 Uniroyal initiated a market development program to introduce a new plastic, one that met our initial research objectives for a high temperature resistant engineering plastic. This new polymer has a heat distortion temperature of 300°F., high impact strength, good processability and is correctly priced. This means it is competitively priced to such engineering plastics as the polysulfones, polycarbonates, and polyphenylene oxides. Specific markets and applications have to be developed since this new polymer does not closely parallel our volume plastics. So far several months of market development work has created a wide-spread interest in this plastic. Product development work has been intensified and we have demonstrated high temperature performance. Experience has helped us to improve on the original excellent high impact strength. Excellent processability is one of the strong points of this plastic. Meanwhile, we had to make a polymer change to resolve an orientation problem encountered while extruding sheet. This helped to reconfirm our policy that we must carefully evaluate any new product, under production conditions, before it is released commercially.

This market development program has demonstrated that we have a product that meets our original research and development objectives. Also, we have gone through the experimental stage when only a limited amount of Arylon was available. We plan to continue the market development phase. We are out to demonstrate performance in specific transportation, appliance, space, and vending machine applications. We are in the important transition period when this new polymer is being turned over to direct sales. Only time will prove whether we really have developed a valuable new product, but we are pleased with the progress to date and optimistic about the future.

Summary

To sell a new product and to make a profit is the ultimate criteria of a successful development project. Competition and rapid change in the chemical industry requires a wide variety of talent to commercialize new technology.

While it is true that a profitable sale is the ultimate test, it is also true that unless a producer has a competitive economic process he probably will not make a desirable profit. A customer will not obtain the properties he expects at the cost required unless this product has good, consistent processability. The financial success to both the producer to make and the customer to use hinges on proper processability. Consequently, it is essential that these characteristics be included in the initial research and development objectives and that they be kept in the proper perspective throughout the life of a project.

The chemical industry has always appreciated the need for new and better products and the need for low cost production. Management has coupled this insight with a willingness to invest substantial amounts of capital in both research and in technically advanced new equipment and facilities.

As the 1960's draw to a close severe competition, international as well as domestic, rapid changes in technology, inflation, and rising costs all contribute to a severe pressure on profits.

For these reasons it is expected that new processing technology will play an ever increasing part in the effort of chemical companies to maintain a record of strong profit growth.

RECEIVED June 6, 1969.

8

Patent and Legal Aspects

THEODORE C. VIRGIL

American Cyanamid Co., Stamford, Conn., 06904

The use of stabilizers to protect plastics against oxidative, infrared, and ultraviolet light degradations presents interesting questions in the area of patents, trademarks, and licensing. With the passage of the Patent Act of 1952, it is now possible to procure meaningful patent protection on the applications of stabilizers to plastics. Significant patent protection to the supplier of stabilizers can now be obtained covering the use of stabilizers in plastics in the form of a method or composition of matter. A trademark which has required widespread consumer acceptance can provide a competitive edge with regard to the sale of the stabilizer and need not be licensed by the owner thereof.

Under the patent system, we think of patents as covering new plastic materials or methods of producing old or new plastics. Actually our patent system can be effective in protecting new ideas relating to the use of additives in plastic materials. The patents which can issue covering such ideas relate to the combination of the plastic with the additive or the use of the additive in plastic material to protect it against degradation such as ultraviolet light, heat, and other deleterious forces. To appreciate how a patent owner may derive benefit from such patent rights, assuming he is interested in more than defensive patenting, it is necessary to understand something about our laws on contributory infringement.

Contributory infringement under our patent laws is incurred by either inducing others to infringe or selling to others a non-staple item which has no substantial non-infringing use and the selling party knows infringement will occur in its use by the purchaser. Contributory infringement is an important doctrine to know in the plastics industry, because frequently we are dealing with situations in which an additive or an ingredient is sold to others who may employ it in a way to infringe either a method patent or a patent covering the combination of the item sold with one or more other ingredients.

Specifically the situation may involve a patent covering a method of reducing ultraviolet light degradation of plastic materials by incorporating into it a certain Compound A, or a patent covering the combination of a plastic material and Compound A. Under the present Patent Law, enacted in 1952, it is now possible for the patent owner who supplies Compound A to others for use in his patented system, to sue other suppliers for contributory infringement. Hence, meaningful patent protection is possible in such cases which could prevent others from selling A in a way as to be contributorily liable.

While the patent owner, who supplies Compound A as described above, may sue others for contributory infringement, and this is possible under Section 271 (d) of our Patent Act, he must observe certain rules to avoid misusing his patent. If misuse occurs, he may not be able to enforce his patent just so long as the misuse exists. The rule to follow is reasonable, namely, as long as the patent owner supplies the Compound A in competition with others, he must make available to all other ultimate users a license on the same terms as he gives to his customers, even though the licensee may wish to buy Compound A from other suppliers.

In the licensing of patents in the plastics field, generally there are certain practices which ought to be observed to avoid difficulties. For example, a licensee should not exercise veto power over who else should be licensed by the patent owner. This is considered a bad practice under our anti-trust laws.

Price fixing of the patented item, while still in doubt in certain limited respects as to how far a licensor or patent owner can go in pursuing the practice, is not recommended at all, because of many possible pitfalls.

While it is a tempting practice because of the obvious commercial gain, if a supplier holds a patent covering the use of Compound A, for him to insist that Compound A should be bought from him to get a license, he should avoid this practice because it constitutes a misuse. Such practice may also involve an antitrust violation in certain situations.

Where a package or plurality of patents are available for licensing, care must also be taken in how the license is offered, because it may be considered a "misuse" for the licensor to insist that all the patents of the package be licensed. The courts frown upon the licensor using coercion to compel the licensing of all his patents.

In recent years it has become very risky practice for a patent owner to license a patent, knowing the patent has serious defects which would lead him to believe it is not valid. Serious anti-trust violations may occur in licensing a patent which the patent owner feels is invalid or even in procuring a patent knowing it will be invalid at the time of its issuance.

The licensor in the plastic field preferably should not license anyone with the understanding that licensee should not sell competitive goods. One can impose on the licensee, where appropriate, and usually in the case of an exclusive license, the obligation to use his best efforts, but not forbid competition in other goods.

It is also important to remember that the courts are refusing to enforce royalty provisions which require a licensee to pay royalty after the licensed patent has expired, if it is the only way covering the goods sold.

Referring again to the plastics business, we can see that it is not only possible to protect a segment of your business through patents covering basic compounds or polymers, but patents covering combinations of materials and uses can provide protection which can be profitable. Admittedly, the practices of old where a patent owner could insist upon all types of privileges are no longer safe or desirable from a legal point of view, however, within the permitted channel of exploitation much benefit can be realized.

The trademark field differs, in that, there is no law which compels the owner of a trademark to license others who are dealing in the same or competitive goods. However, a problem can arise if the trademark owner will insist on his licensee buying from him the goods to which the mark will be affixed, if the goods are available from another source.

RECEIVED, June 20, 1969

9

Application Development

N. BRUCE DUFFETT

Marbon Division of Borg-Warner Corporation, Washington, W. Va. 26181

In the development of any market with a new or modified engineering thermoplastic the first step involves careful and thorough planning. The plans should consider the nature of the market to be sold. What are the needs of the market in terms of product definition, technical service requirements. type, location, and size of customers, and an analysis of your own capabilities to satisfy the needs of the market. With a thorough understanding of the market it is possible to establish dynamic objectives. Each objective must not only be quantified but include a time schedule. Four case studies illustrate the preliminary planning and the establishment of objectives. Each of these case studies is in various stages of completion and together serve to demonstrate the problems and complexity of market development of an engineering thermoplastic.

What is the key to market development? Can anyone unlock the secret? If so, how do they do it? I wish that I could offer you some assured path to success. This cannot be done. If these comments can stimulate some thoughts—present a few ideas—then they will accomplish the mission.

With what does market development concern itself? It concerns the future. Charles Kettering of GM was credited with saying "We should all be interested in the future because we expect to spend the rest of our lives there."

Karl Taylor Compton—scientist and educator—wrote, "While it is occasionally pleasant to think back, it is far more profitable and interesting to think ahead. Adventure, progress, and exhilaration of achievement always lie in the future, and their planning should be the chief concern of the present." Compton must have been referring to market development when he used words like adventure, progress, and exhilaration of achievement, because each of those is very much part of this effort.

There are several steps which we consider essential to successful market development. First, we should describe as clearly as possible the nature of our business. Is it philanthropic, or do we intend to make a profit? Does it contribute to another function or must it stand alone? How much is known about the business we are attempting? Who are the competitors, and what are their resources? How about our own capabilities? Can they be defined accurately? What is our knowledge of the marketplace? Are the sources reliable? What assumptions are being made? Are they accurate? These and other questions should be asked and answered to organize our current position.

From here we establish a set of objectives and try to make them dynamic. A dynamic objective says what is to be done by when. These can be classified as standard objectives, problem-solving and innovative. In market development the innovative objective is always the most challenging and difficult to achieve. These are the problems. – In a book called "Grooks," Piet Hein wrote–"Problems worthy of attack prove their worth by hitting back,"(1).

Once the objectives have been established, then it is necessary to develop programs or projects to support each objective. Describe the program in detail, indicate its purpose, define responsibility, establish a schedule, and estimate the money, manpower, and material requirements.

There you have the essence–in a few words–of a sound approach to market development. But it lacks one essential characteristic. Let's call that motivation. Motivation comes only from a complete understanding of the product, its value to the customer, and a personal commitment to the entire project. This is known as justification and summarizes the reasons for undertaking the market development project in the first place. It is the keystone that holds the arch in place and provides the basis for commitment to a successful venture.

Camper Unit

Today let us consider four case studies involving market development activity. The first of these is the application of an ABS plastic in the shape of a thermoformed sheet to create the entire structure of a camper unit mounted on the back of a pickup truck. Less than three years ago, William Suiter, then President of Marbon Chemical, and representatives of Ford Motor Company agreed that a thermoformed camper body offered unique design possibilities and economics in production. They agreed that an ABS plastic sheet would meet the most rigid specification requirements. They understood from research reports that recreational vehicles of this type constituted one of the fastest growing markets in America. At the time there was available a thermoforming machine with the capability of handling a 10 feet x 25 feet plastic sheet although no tools of this magnitude had ever been tried before. Various designs were made and tooling costs estimated. It was determined that the tooling costs could be amortized over the first 10,000 units and annual requirements of 10,000-40,000 were discussed. The current market for this type of recreational vehicle is in excess of 120,000 units, representing a market potential of 40 million pounds of ABS sheet.

In March of 1966 the custom camper project got underway with a five-stage plan to design and build a prototype. This involved a full-scale clay model, preliminary tooling, foaming jigs, tests, evaluations, and modifications all prior to the final tooling. Design called for a unit split down the center vertically, so this meant a right half and left half inner and outer shell—four massive tools. The inner and outer skins were bonded together in jigs with an insulating polyurethane foam. The plan extended over a six-month period to prepare a fully completed prototype. Once the prototype was approved by Ford, a purchase order was issued for 10,000 units and production tooling was initiated. Many scheduling and production problems were encountered and solved. Production was underway in the Spring of 1967. However, everything was kept under wraps until the first week of April 1967 when the Goldline Camper was presented to the public at a press meeting in Carefree, Arizona. A number of bulletins had been prepared. Advertising and publicity releases were scheduled by both Ford Motor Co. and Marbon Chemical. A major breakthrough had been accomplished in just one year of coordinated effort between plastic raw material suppliers, design engineers, large sheet extrusion and thermoforming facilities, tool and die makers, and an automotive supplier. A market need had been recognized and supplied. Many individuals contributed willingly of time and talent to make the Goldline Camper a reality. Many

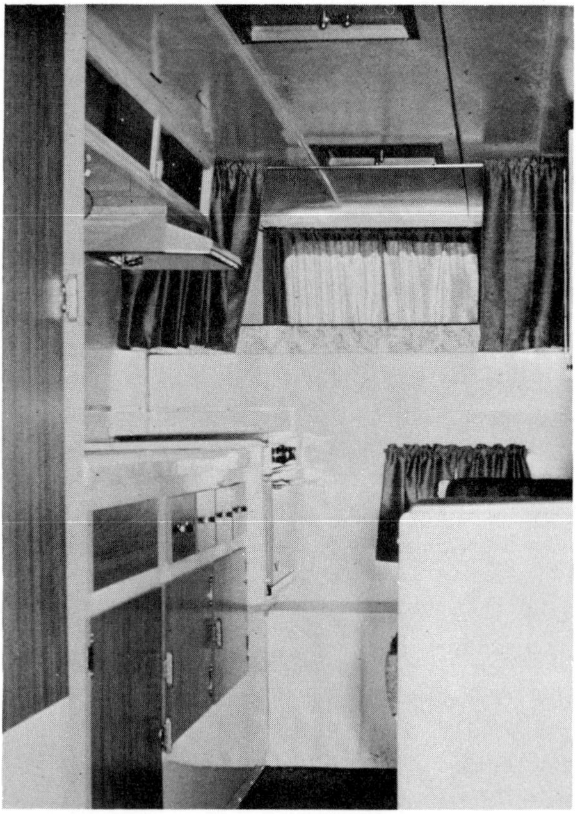

plans were developed but the ultimate success of this development can be attributed to the belief and commitment on the part of courageous and creative individuals.

Furniture Industry

Our second case study concerns the application of expanded or foamed ABS to the furniture industry. During the early part of 1967, Marbon Laboratories developed a product that could be rotationally cast to the approximate density of hardwood with strength, chemical, and dimensional properties quite superior to wood. The initial capability analysis revealed that this product could out perform wood by a wide margin when applied to curved surfaces in the furniture field, but we knew very little about the furniture industry. So we sought professional guidance by employing a consultant with experience in this area. He performed a valuable service and supplied designs for a variety of chairs from which one was selected and a model prepared. The model shell was evaluated, tested, and used to make a two-piece matched aluminum casting mold. This mold was used to produce several shells which could be shown to a furniture manufacturer to determine his interest in the concept. In this instance, Thayer Coggin in High Point, N. C. was the furniture manufacturer and his designer, Milo Baughman, participated in some of the earlier discussions. The shell was produced at a cost of less than $20 compared with the same shell produced in wood at a cost of $38. The basic shell was upholstered in two different patterns using many different fabrics and soft vinyl skin on top of urethane foam padding. The unit was mounted on a variety of bases to create a line of novel and unique furniture. The curved shell was also divided in half and used to make a love seat to match the original

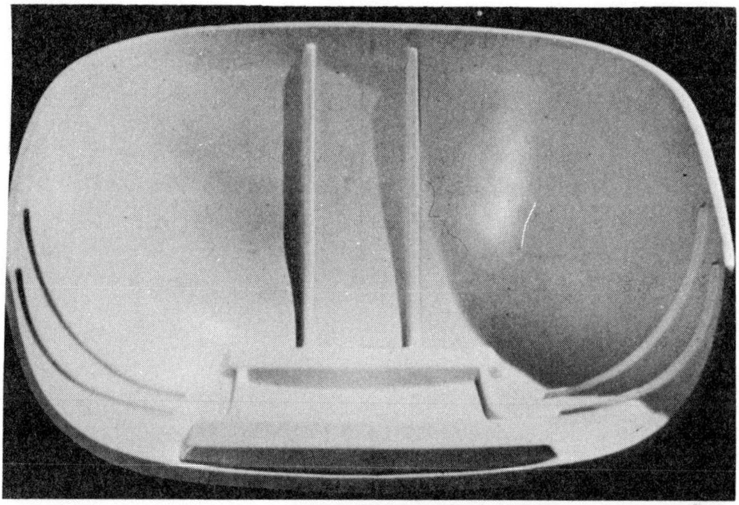

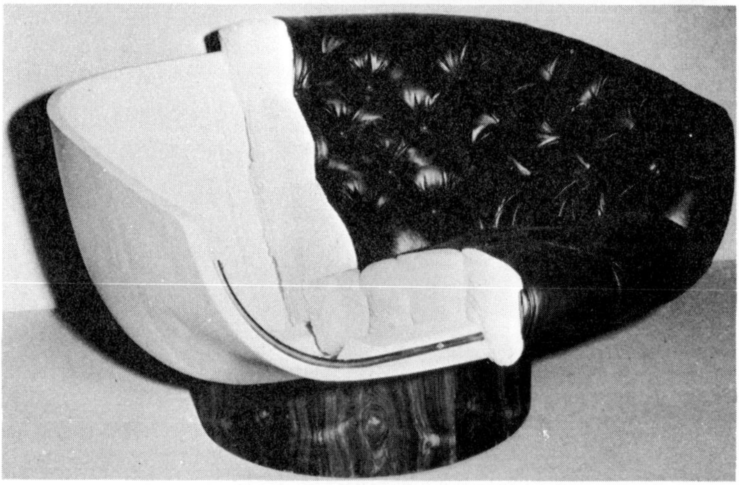

chair. The chair and love seat were introduced at the October '67 furniture shown in High Point, N. C. where they received an enthusiastic reception from the trade. Most important a new concept in the manufacture of furniture had been demonstrated. This concept introduced a new material and a new manufacturing technique to an old line industry with the result of lower cost and more efficient methods. Extensive advertising in the New York Times by Macy's and Abraham Strauss in New York were part of the total program. Since this original effort, many furniture manufacturers are evaluating this technique and applying it to their line of products.

Electroplating Plastics

Electroplating of plastics offers an opportunity for us to review a third market development effort. In the early days of plastics, people were successful in depositing a metallic surface—generally chromium—by vacuum metallizing. This was also known as encapsulation. The appearance was good but the adhesion was negligible. Electrodeposition or electroplating seemed to be the only answer to the lack of adhesion and to do this, the surface of the plastic part must be made electrically conductive. Laboratory efforts over several years developed a modification of ABS resin which could be etched in such a manner that it could be coated with an electroless nickel to prepare it for the standard electroplating bath. We were marketing the ABS plastic Cycolac for this purpose through '64 and '65 but platers were continually plagued with problems of the various chemical baths. It was recognized that the plater's need could be satisfied by marketing a complete system—not only the plastic—but also the cleaner, etchant, catalyst, neutralizer, and nickel to provide the plater with one source of supply. The preplate system could be

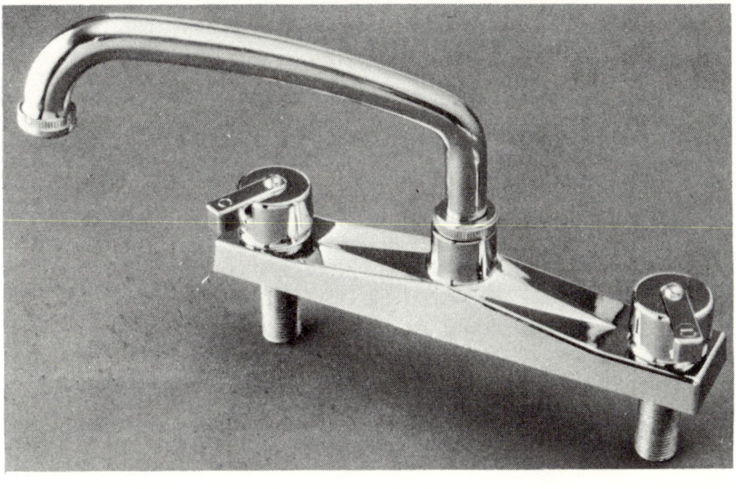

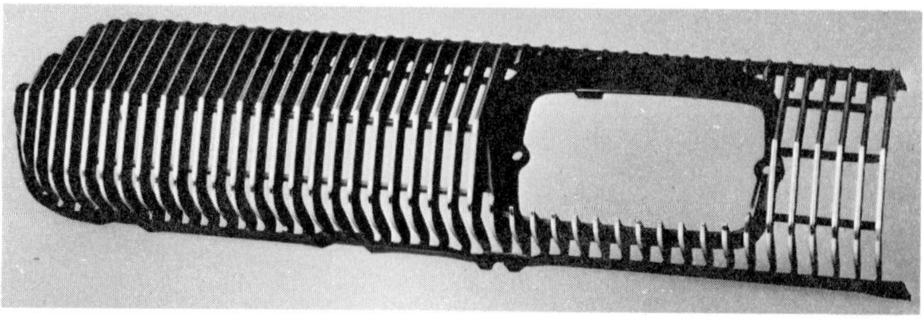

designed directly for the plastic involved. The concept was good, the need was identified, and the product received an eager reception. It permitted the plater to improve performance, lower rejects, and actually reduce preplate costs to less than half of those considered normal for electroplating. The Marbon preplate system was first sold in January 1967 and in two years has witnessed a rapid growth amounting to a 40% penetration into the marketplace. A high degree of competence and extensive technical service were factors that had to accompany the product. Today many of the parts of your automobile that you assume to be metal because they have a chromium surface plate are in reality plated plastic parts. Up to the present time the plating of plastics has been almost entirely decorative, but recently the industry has seen some functional parts. Plastic moldings are coated with metal to impart a surface property needed in the product. This opens up a vast panorama for the future. Market developers can now look at many metal parts and ask the question, "does it have to be metal all the way through?"

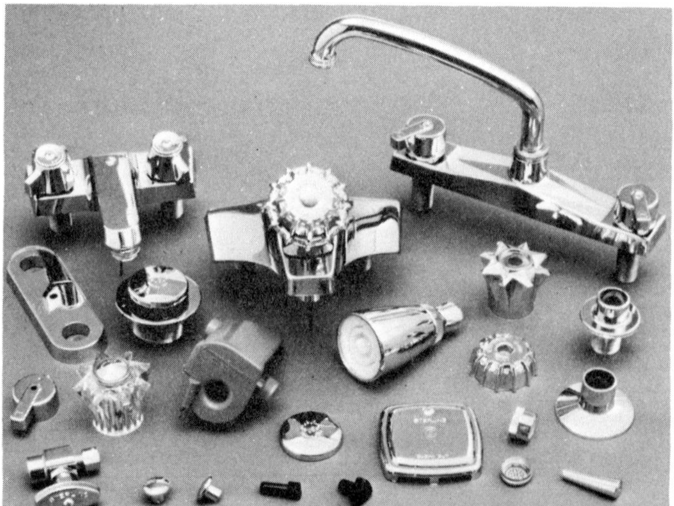

Environmental Control

Our last study takes us away from the plastics industry into an entirely new realm, that of environmental control. Man, through recent decades, has learned how to control the temperature and humidity in confined spaces. Witness the vast growth of the air conditioning industry over the past twenty-five years. Now he has learned how to destroy odors. Why not live in an odor-free atmosphere as well as one of controlled temperature and humidity? We can—and the product to accomplish this is known as Purafil—a purple pellet made from activated alumina and potassium perman-

ganate. Here, the market developer is confronted with an education problem. How does one teach the public or even the consulting engineer to understand the concept of odor control? How does one even measure the potential? It is not a short term problem and patience is mandatory. You might inquire where is the market need. Initially we felt the major market was residential. In 1966 a large scale effort was directed in Houston, Texas to reach the residential market for odor control. The effectiveness of various advertising media to produce awareness and to produce sales was measured. A program involving TV, radio, billboards, newspapers, and magazines was carried out. Although we were able to determine the effectiveness of the various media we also learned that the major market for odor control at present is not the residential market today. Other efforts have been directed toward odor problem centers such as hospitals, autopsy rooms, beauty parlors, allergenic clinics, bars, restaurants, public spaces such as auditoriums and exhibition centers. In control of odors, Purafil oxidizes gaseous pollutants. It is particularly effective in low concentrations. It has been found that it will protect sensitive electronic gear from oxidation by SO_2 which is below the capability of the human nose to detect. It also is quite effective in control of H_2S or C_2HO. These properties make it particularly attractive in the protection of computer centers for oil refineries, steel mills, paper mills, etc. So the prime market development effort for Purafil today is directed toward public spaces and removal of gaseous pollutants where they would adversely affect the life of electronic gear.

Success can be accomplished through the combined efforts of many people – those informed and able sales representatives, those air conditioning distributors and dealers, those original equipment engineers who have the vision of the future, those advertising and public relations men who know how to reach the public, those in management who have courage and confidence in the future. This development program is underway now. The final chapter has yet to be written. Someday soon we hope you will hear more about odor control and its place in total environmental control.

The Factor of Time

All of these cases considered have not been short term efforts. Market development requires at times infinite patience and understanding. The Goldline Camper from concept to commercial reality took just over one year. By comparison this is a very short time. The application of foamed ABS to the furniture industry is still in the process of commercial development. It has been in this process now for more than two years and will probably require one to two years more before it can be considered a commercial success. The study of electroplating on plastics has been under development for a period of at least seven years. It can now be considered to be a truly commercial product. Purafil, however, was discovered about ten years ago. It was first marketed by our Ingersoll Products Division as part of a line of hospital supplies. The next effort was an exclusive marketing arrangement by York Division as a filter for air condition-

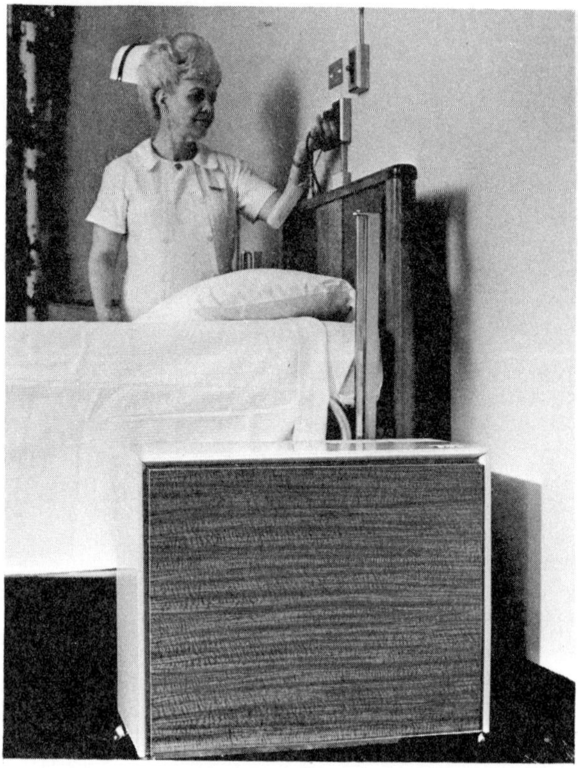

ing equipment. We in the Marbon Division have been concerned with marketing Purafil now for somewhat more than three years. So all developments do not occur rapidly. Some do take time. Our poet, Piet Hein wrote about TTT.

Put up in a place
Where its easy to see
The cryptic admonishment
TTT
When you feel how depressingly
Slowly you climb
Its well to remember that
Things take time.

Literature Cited

(1) Hein, Piet, "Grooks," M.I.T. Press, Cambridge, Mass. (1966).

RECEIVED June 6, 1969

10

Chemical Marketing and Economics

WILLIAM P. MASTROLIA

Celanese Chemicals Group Company, New York New York 10036

Relative to the large volume commodity-type plastics, engineering plastics are characterized by fairly specific physical properties, proprietary technology, low volume, high investment cost per pound of capacity, and high average selling price. The successful marketer of engineering plastics must first determine potential market opportunities based on physical and cost properties. Next market research and market development functions must home in on specific customers, their potential, specifying influences, purchasing practices, customer fabricators, government and other regulatory influences. Competent technical support is required for the sales force. This includes plant, technical, application, and product development labs and field technical service. The stockholders' return on the highly expensive development and support activities comes after the sales department brings these activities to a climax in the form of orders.

Success in any business enterprise does not come easily. The polymer and plastics industry, demanding of large technical expenditures to invent such products, requiring substantial capital investment for plant and facilities to produce them, and above average costs to bring them to market, increases the risk a company takes when a decision to produce engineering plastics is made. Payoff on the total investment of expense and capital comes when product in sufficient volume and reasonable selling price is moving out the door.

This paper will cover the problems plastics producers encounter when they market their products and attempt to obtain a satisfactory return on the stockholders' investment. Explaining the former is almost as difficult as accomplishing the latter. I would like to approach the problem first from the viewpoint of the material supplier and the end user. I'll then try to show how buyer and seller get together with several case histories illustrating these points.

The Problems of the Seller

Selling is one aspect of marketing which in itself comprises a number of separate functions. Market research, market development, applications development, technical service, and advertising and sales promotion are five of these marketing functions.

Market Research

How do each of these functions help the salesmen to sell? Prior to making the decision to produce a new product based on a company's proprietary technology or to produce a product currently on the market based on self-developed or licensed technology, management should have considerable detail on market size, end use industries and applications, prices, competition, possible market share under varying conditions, product mix requirements, and a list of prospective customers. Forecasts for each of these factors should be made for three, five, and possibly ten year time spans. Incidentally, this type of information is much more difficult to develop for new products than for established products.

All of this information is obviously essential for project evaluation to determine if the company belongs in the business. But is it of any value to the sales manager and product manager when sales plans must be made by industry, by end use, district, territory and customer? Presumably market research has identified volume and volume growth rates from the known profile of the products physical and cost properties. To be of assistance in developing the sales plan, however, it is necessary now to identify in very precise terms the volume potential in specific companies, for specific applications, over a one or two year time span.

Uncovering this information for an established product is reasonably routine using conventional market research techniques. The results obtained will be directly proportional to the effort input since there is no way that I know of to extrapolate to a universe from a research sample. After the information is gathered, it would be normal to turn the product over to the next marketing activity for the development effort required prior to the sale.

Using market research techniques for a new product as yet unknown to potential users is more apt to result in failure than success. End users, fabricators, and other purchasing influences generally respond objectively and frankly when questioned on the potential value of a new product. On a second interview, however, it would be difficult to obtain more definitive statements on precise potential by year for volume, price, package type, order points, and other information required to run a business. To develop this information for a totally new product requires a second level of marketing activity.

Market Development and Applications Laboratory

Market development is another marketing function known by many names in many companies. I view it as that department which contacts the prospective customers identified by market research or other source data. They attempt to sample material for part evaluation, in the case of established products, or to arrange to build prototypes in the case of new products.

Because of the nature of their activity, market development must work closely with the applications laboratory. Applications is a group of specialists who develop detailed technical information on the properties of the new plastic raw material. They determine the functional suitability of the plastic part after it has been produced from the customer's mold or of the prototype part built by the lab group. This testing of course, includes the usual chemical and mechanical analysis as well as environmental testing under the conditions in which the customer intends to use the part.

Other important functions of the lab should include:

Knowledge of how the plastic responds to all types of processing and fabricating equipment.

How to make the material run under field conditions

Mold design

Part design

Material behavior under a wide range of physical and chemical environments, and under varying stress loads at a range of temperatures.

The applications lab should be the center of product information. It should be staffed with people who not only carry out these functions, but also communicate the results with customers. For this reason a close relationship between market development and the applications lab is absolutely essential for a meaningful, coordinated approach to the customers, for more often than not the period between market development's first call and repeat order business will span two to three years.

Advertising and Sales Promotion

The identification of specific market opportunities and the development of customer awareness of your product and your desire for his business, is aided immeasurably by advertising and sales promotion. Advertising programs can serve a number of objectives. Target markets can be probed to determine areas of optimum interest. A communication link can be established with key people within these markets. Once advertising screens readership surveys and inquiry responses and sifts out the desired audience, it is possible to plan a campaign whose objective is to communicate your message to the decision makers in the target markets.

For example, Celanese announced Celcon acetal copolymer in February 1961. By the last half of 1962 and the first half of 1963 we carried through on a specific advertising and promotion program which consisted of:

1. Three direct mail promotions.
2. A series of five ads aimed at:
   (a) Design Engineers
   (b) Plastic Molders
   (c) Screw Machine Operators
   (d) A broad Industrial Audience
3. A sales presentation flip chart.
4. Major exhibits at the Design Engineering and Plastic Industry Trade Shows.
5. Customer Seminars.
6. Case history publicity and public relations back-up.

The ads, seminars, and case histories were all designed to illustrate a particular point or subject known to be of major interest to people broadly defined as influencing the purchasing decision. These purchasing influences vary by number, type, interest and function, depending on the company and the market the company operates in. Design and development engineers are generally interested in properties and performance. Purchasing is interested in availability, reliability, consistency, packaging, and price. Management is broadly interested in value of the end product and the profit the use of this new plastic will ultimately generate. The molder is interested in consistent raw material quality, forming ease, machine throughput, and price.

The case history approach was also used to interest design engineers in end use companies by highlighting successful applications of Celcon. The case histories spelled out reasons why the material was selected over another plastic or a metal and emphasized performance requirements of the material in the application.

In summary, advertising can represent an efficient and economical method to prospect for business and to find areas of profitable interest. It must be coordinated, however, with other functions in marketing for optimum results.

The Problems of the Buyer

Generally two people are required to complete a sales agreement—the seller and the buyer. So far my remarks have focused on the problems of the material supplier—the seller. What of the buyer? What are his problems? Why is he so frequently difficult to comprehend and to deal with? Let's look at some of his problems.

In the Thirties and Forties there were no more than four or five plastic materials produced in relatively low volume and sold at relatively high prices.

Today there are between 45 and 50 plastic materials. Physical volume, averaging a 13% per year growth rate for the past ten years, reached an estimated 16 billion pounds in 1968, higher than that of any metal except iron and steel, and approaching the total for non-ferrous metals. The number of formulas, grades, and types of these materials is greatly expanded by the use of plasticizers, fillers, and polymerization alternatives. All of these formulations are presumably different from one another and offer the user a broad material selection to fit his property and cost requirements.

Consider also that almost every plastic raw material is made by more than one producer. For example there are now about seven producers of ABS, five producers of methyl methacrylate, eight producers of polypropylene, four producers of fluorocarbons, fifteen producers of polyesters, five producers of nylon, twenty-one producers of polystyrene, thirteen producers of low density polyethylene, eleven producers of high density polyethylene, two polycarbonate producers, twenty-four PVC producers, three epoxy producers, two acetal producers, three cellulosics producers, and thirteen producers of phenolics.

Although engineering plastics in the topic of this book, I would maintain that the end user must have broad product interests spanning the offerings of the entire industry. I would also maintain that while some of us close to the trees may view an engineering plastic as one having some balance of properties which are different from products like polyethylene or polystyrene and PVC, the end user doesn't necessarily have the same pair of glasses.

Within industries that consume large quantities of all plastics such as automotive, appliances, business machines and the broad industrial category, there are literally hundreds of applications for the so-called "non-engineering" plastics. True, these applications may not require the broad balance of properties achievable with the engineering types, but as far as the design engineer is concerned these materials do the job he wants done at the right cost. He could not serve his company's best interest if he did not know of their availability and keep up with the almost daily change in product variations.

The significance of these comments then is that the end user, his product design and development engineers, purchasing and management people are subjected to continual barrages from raw material suppliers in the form of requests for: market research information; the establishment of market and application development programs; and advertising and promotional campaigns from this large list of suppliers of a large number of plastic raw materials.

The scope of the end user's problem is further magnified by the observation that his other suppliers such as the metals, rubber, and other construction materials people are not standing still. They too are making progress and are anxious to keep him informed.

It should be fairly obvious, then, that because of the tremendous volume of information, the problem of keeping informed is a difficult one. The raw material suppliers through various techniques can keep the buyer or end user current with product developments. The end user's problem is to stay current while still performing the job he's paid to do. It's a difficult one obviously, involving compromises which are often made against the raw material supplier who does not measure up to the standards established.

Case Histories

At this point, I hope you have an overall view of the problems of the buyer and the seller and how the relationship is established and nurtured. It would be presumptuous of me to try to explain in any kind of detail how a sale is made. There are far too many variables in each situation to generalize or draw conclusions. I would like to report on several case histories at Celanese which I believe illustrate some of the problems we're discussing today.

The first involves an attempt to replace a competitive plastic in an established application. Market research and market development had reported that a nylon material was being used to make tubing for such items as automotive brake cables. The market was estimated to be several million pounds and concentrated in only a few companies who were making the cables and selling directly to the end user.

Based on knowledge of relative properties and costs, it appeared to us that Celcon had some advantages to offer in the application. We quickly learned that people making tubing from nylon were extruding at rates up to 150 feet per minute. At the time, our company had very little experience with the extrusion properties of Celcon and we learned that under conditions which were being used to make nylon tubing our material extruded at only 50 to 75 feet per minute, so low as to offset the cost advantages of Celcon.

The potential volume involved and our assessment of the property benefit of Celcon in the application were sufficient to justify a lab program whose objective was to develop a polymer or modify the extrusion process so that Celcon would extrude at comparable or faster speeds than the material being used. After some nine months of effort an extrusion die was designed that permitted extrusion speeds of 450 feet per minute while making tubing with the tolerance specifications established by the customer. The foregoing is an example of a type of process development work described in the preceding paper.

The contribution made by Celanese was the development of a processing technique to enable the end user to convert to a raw material lower in cost than his present material. This new material was capable of producing an end product which matched or exceeded all specifications at a lower production cost. As a result of this, we obtained most of the business which we continue to enjoy today.

A second example involves the development of a completely new application. Again market research had identified a plastic potential in the plumbing industry as a replacement for die cast brass for functional parts which were not chrome plated. At the time Celcon was a new product and the company had only limited experience with the design characteristics of the material with respect to wall thickness, thread design, and long-term environmental behavior for this end use. In an attempt to develop this information, a joint program with a manufacturer of plumbing equipment was established. The manufacturer's role was to provide the basic requirements and standards of the industry in a broad base of applications. The role of Celanese was to build prototypes employing our knowledge of the material and its fabrication which would meet the requirements for plumbing applications so that environmental testing could be conducted. As a result of the joint effort, the answers to many questions were uncovered although the industry did not yet adopt plastics on the same basis as the metals. It should also be noted that a useful life of up to 20 years is expected for materials used in construction type applications resulting in some conservative practices by the supply industries.

At this point the information accumulated over the years was presented to the Celanese affiliates in Germany, Japan, and the U. K. They were able to develop the use of Celcon for hot and cold water faucets. As a result of this experience there has been renewed interest shown by domestic manufacturers. The problem of developing totally new applications for engineering plastics should center on overcoming the natural reluctance to change. This is probably one of the most difficult chores faced by the marketing department of companies making engineering plastics but should also be the most rewarding because of the potentially large volume.

On occasion a potential market for plastics has been clearly defined and continually tantalizes resin suppliers by always remaining just out of reach. Such was the case with the dairy industry and the plastic milk bottle.

Although high-density polyethylene is not generally regarded as an engineering resin, largely because of its high volume of consumption, I suspect many of its applications call for a threshold level of engineering performance that is well above normal product specifications. High-density polyethylene pipe for gas service and distribution lines and polyethylene wire and cable insulation are good examples of these engineering requirements.

Milk bottles fall into this engineering category because of the precision requirements for filling and handling on high speed automatic equipment. The milk bottles must also have the structural integrity to withstand the high temperatures of cleaning and sterilizing equipment. In addition, they must meet minimum standards of the Food and Drug Administration as to taste and odor.

Since the early 1960's many of the largest resin suppliers have seen the huge potential volume to be gained by the replacement of the glass milk bottle with one made from a lightweight plastic. Several invested heavily in dairies and dairy equipment manufacturing without gaining acceptance by either the consumers or the dairy market.

Where other larger companies had failed to break into the milk bottle market, several small dairy supply entrepreneurs in Pennsylvania and the Southeastern region of the U.S. were able to generate consumer acceptance in local areas. These dairy supply houses turned their hand to blow-molding gallon and half-gallon polyethylene bottles and supplying them to local dairies largely for use in retail stores. When this market took shape, we moved to develop a resin that would be especially suited to the type of blow-molding machinery these dairy supply people were using. Within a short time our laboratories came up with a specially engineered resin that would meet all the necessary stringent specifications of the dairy industry and also permit blow-molding of bottles on shorter and more economical cycles. Consistent lot-to-lot product uniformity played a large part in establishing this new resin in this market. At present

Celanese enjoys the major share of the merchant plastic milk bottles business in the United States because of this engineered high-density polyethylene resin.

The pattern then is fairly predictable. First is the discovery of the need. This is usually accomplished by market research and market development but more often by the company's field sales force, one of whose objectives should be the constant search for the need.

Next the identified need and the properties of the engineering plastic have to be meshed. These properties include physical, mechanical, environmental, processing, finishing, and cost characteristics. The way these properties stack up and perform under the use conditions of the part relative to a competitive material really determines if a sale can be closed or the development continues. Making all of this gel takes time, and requires close cooperation between the raw material supplier, end user, and the fabricator.

Finally, your product must do something for the customer. It must have some clear advantage over the competition and the customer must be convinced that this advantage will last over the useful life of the finished product. To accomplish these tasks, to move the product in sufficient volume and at a satisfactory price depends on a highly coordinated effort by the entire marketing staff.

RECEIVED June 6, 1969

11

Mill Shapes for Engineering Application

THOMAS D. TREMMEL

Cadillac Plastic and Chemical Company, Detroit, Mich. 48232

This article traces the rapid growth of the family of plastic materials as engineering products. Increasing numbers of diverse materials coupled with improved machining techniques increases daily the broad applications base for mechnical plastics. A gradual annual price decline of materials augmented by wide spread product distribution and a systems approach to product development and marketing have greatly extended their multimarket base. Realistic cost vs. performance pricing of finished machined components has successfully eroded many traditionally metals oriented markets.

Currently available to the marketplace are 40 or more different thermoplastic and thermosetting plastics materials. This "embarrassment of thermoplastic design riches" makes increasingly difficult and laborious the proper implementations and proper use of these myriad materials. This is compounded by an availability factor. Not only is industry harassed by what material or materials to use, but how to use them. There exists today in design and engineering circles a genuine poverty of information concerning the proper use and application of the specific beneficial properties of the huge category or family of engineering materials we call "Plastics."

Industry estimates indicate that up to 5% of the total resin production finds its way into prototype or mill shape plastic products. By "mill shapes" is meant those primary uniform configuration subject to established cross-sectional and length tolerances. While this estimate is necessarily conjectural, the best available information indicates that this range is accurate. *Modern Plastics* magazine estimated mill shape production for 1968 in acrylics, cellulose, nylon, acetal, polycarbonate, high density polyethylene, polypropylene, poly(vinyl chloride), and copolymers to approach 336.4 million pounds. Total United States resin production for 1968 slightly exceeded 16 billion pounds.

Much of the mill shape production is converted in house by the processors themselves—such as in the case of polyethelene film for agricultural uses which is 100 million pounds, and PVC swimming pool liners, etc.—there exist large markets of the mechanical plastics.

The mechanical plastics we refer to as:

Polyimides	Acetals
Chlorinated Polyethers	Polycarbonates
Polyphenylene Oxide	Acrylics
Nylons – Natural and Filled	Polyolefins
Fluorocarbons – Virgin and Filled	Polysulfone

By definition these are the materials whose physical properties and cost performance ratios allow them to compete with and replace traditionally accepted ferrous and non-ferrous metals as well as other mechanically functionable design materials.

After a material section has been made for a specific application, the production mode decision must be considered. Generally speaking, the part configuration and the quantity of parts to be produced will dictate the most efficient production process.

All of the mechanical plastics we have here discussed can be machined. Most are worked with little difficulty. The mechanical plastics are being used in an almost unlimited number of applications and virtually every type of industry. Machining has been accepted as the most efficient method of producing prototypes and parts in small or medium quantities. Within the last several years, machining has been recognized as a highly efficient method for many production runs in the tens of thousands, hundreds of thousands, and more unit quantities. Heretoforé injection molding has been looked upon as the least expensive method of producing parts of more than 10,000 unit quantities. This today is not always so. Fabrication technique depends on product properties and shape configuration required.

Machinists have learned through experience, and are certainly well aware, that machining plastics is profitable. While machining plastics is not particularly burdensome, there are specific considerations that must be followed when machining plastics. Because plastics have a much lower deflection temperature than metals if too great a heat build up is allowed, the plastic may tend to gum tooling. Plastics are more resilient than metals when close tolerance is required. The operator must compensate for the resilience effect of plastic.

Machining Criteria For Mechanical Plastics

Good shop practices should be followed when machining plastics. Tools should be kept sharp and a system of efficient chip removal should be employed. Chips can usually be removed by air jet, a

Table I. Properties of Materials for Low Friction Applications

	Cost, Base Resin (¢/in³)	*Abrasion Resistance (mg. loss/1000 cycles)*	*Flexural Modulus (10^5 p.s.i.)*	*Apparent Modulus (10^5 p.s.i.)*	*PVRating, Dry Continuous (X 100)*
Fluorocarbons[a] TFE	25 to 40	7	..	1 0.4	to 2.5
FEP[b]	43	13.2	0.95	0.6 H	to 0.9
TFE Fabric[c]	106	..	..	..	5 to 50
Filled TFE[d]	26 to 58	8 to 26	1.2 to 2	0.6 to 1	5 to 35
Nylons[e]	3.7	6 to 8	1.5 to 4	1	2 to 3
Acetals[f]	3.3	6 to 20	3.1 to 4.1	2	2 to 3
Acetals[g] TFE-Fiber	26.5	..	5.51	2	7.5
Polyethylenes[h] High Density	0.9	6	1.3 to 3.2	1.5	..

[a] Usable temperature range -430° to +550°F., does not adhere to tacky materials; nongalling, absorbs abrasive particles, chemically inert.

[b] Readily injection molded and extruded, does not adhere to tacky materials; chemically inert.

[c] High load capacities at low speeds (50,000 p.s.i. at 0 f.p.m. no set under heavy static loads, requires low clearance. Not recommended for over 200 f.p.m. rarely used over 50 f.p.m.

Comments on Table I.

Properties Required: Low coefficient of friction, even when nonlubricated. High resistance to abrasion. Fair to good form stability, heat resistance, and corrosion resistance.

Suitable Plastics: Fluorcocarbons (TFE and FEP), filled fluorocarbons (TFE). TFE fabrics, nylons, acetals, TFE-filled acetals and high-density polyethylenes.

Other suitable Materials: Babbitts, bronzes, cast irons, prelubricated woods, graphite and cements.

Consider Plastics When:

1. Corrosives or abrasives are present.
2. Lubrication might contaiminate product being processed.

(Bearings, Bushings, Slides, Guides, Valve Liners, Wear Surface)

Heat, Distortion Temp. at 66 p.s.i.	*Thermal Conductivity (Btu-in/hr-ftt^2 °F.)*	*Thermal Expansion per °F.) X10^5*	*Water Absorption in 24 hr. (%)*	*Slip-Stick*	*Coefficient of Fiction*	
					Dry	*Lubricated*
250	1.7	5.5	None	No	0.04	0.04
<250	1.4	4.5	<0.01	..	0.08	0.08
..	1.7	8	None	0.02 No	0.02 to 0.25	0.02 to 0.25
<250	1.7 to 20	3 to 9.7	None	No	0.16 to 0.28	0.06
340 to 360	1.4 to 2	4.6 to 7.1	0.4 to 3.3	Yes	0.15 to 0.40	0.06
316 to 338	1.6 to 1.9	4.5 to 6	0.12 to 0.41	No	0.15 to 0.35	0.1
329	1.7	4.6	0.6	No	0.12	0.07
140 to 182	3.4	6.5 to 16.7	<0.01	Yes	0.21	0.1

[d] High load capacities.

[e] Absorbs and engulfs abrasive particles, nongalling available in massive shapes.

[f] Stiffest unfilled thermoplastic, creep resistant.

[g] High creep resistance, good wear resistance, most efficient at high loads & low speeds.

[h] Nongalling.

[i] This table was taken from *Plastics Quarterly*, *Machine Design.* (Dec. 12, 1968).

3. Assembly must operate above or below useful temperature of conventional lubricants.
4. Maintenance-free operation is desirable.
5. Complex lubrication systems would otherwise be required.
6. Weight is a major consideration.
7. Electrical insulation must be provided.
8. Noise must be controlled.
9. Galling and scoring must be controlled.
10. Galling and scoring must be minimized.
11. High-load, low-speed operation would squeeze out conventional lubricants.

Comments on Table I (Continued)

12. Slip-stick characteristics would be objectional.

Consider Other Materials When:

1. Temperatures over 500°F. are encountered.
2. Heavy radial or thrust loads exist.
3. Continuous high-speed operation is required.
4. Shaft deflections must be minimal over long periods of time.
5. Wear on shaft is preferable to bearing wear.

Consider Plastics-Other Material Combinations When:

1. Maximum heat dissipation is required.
2. Cold flow must be minimized.
3. Loadings too high for solid plastics must be withstood.

Property Summary: Nylon recommended for general-purpose bearings and wear surfaces. Fluorocarbons (especially TFE) for sliding or low-speed-rotating dry bearings, for highly corrosive service, or service in extreme temperatures (-430 to +500 F). Acetals for submerged or humid service, and when resistance to creep is important. Acetals and fluorocarbons for valve liners or slides to eliminate jerky starts and slip-stick. Filled Fluorocarbons for heavier loadings and high creep resistance. TFE-filled acetals for heavy duty sleeve or sliding bearings. TFE fabric for ball-and-socket and thrust bearings; sliding bearings under heavy load, low speed. High-density polyethylene for lowest cost at very low speeds and loads. Fluorocarbons for nonstick surfaces.

---

combination of water soluble oil and water, vapor mists, submerision, vacuum or by use of a chip breaker or mechanical deflection. Plastics are efficiently machined by high speed steel tools. Cutting edges must be kept sharp and must be dubbed after sharpening. Dubbing is particularly important with acrylics. Other plastics may dub the cutting edge on first contact with the material. Generally speaking, tools should be set with a zero or slightly negative rake and should have a scraping rather than a cutting action. Care should always be taken when selecting speeds and feeds for the more brittle plastics such as acrylics. Correct machining techniques produce a high quality cut with a minimum of induced stresses in the work. Some plastic materials, polycarbonate particularly, should be annealed following the machining operation and in some cases prior to the machining. While annealing is not absolutely required for most jobs, it does help to relieve stresses built up during the machining. A great majority of the thermoplastics can be machined dry. Cooling by air jet, vapor mists, or a solution of between 10 and 20% water soluble oil and water is required at higher speeds. Cooling compensates for the lower heat distortion temperature plastics. The higher resilience of plastics also may require drilling oversized holes. Polycarbonate, polyphenylene oxide, nylon, and acetal are generally machined similarly. Fluorocarbons and polyolefins (polypropylene and polyethylene) fall into a similar category. High density and ultra high molecular weight polyethylene machine much easier than lower molecular weight and low density polyethylene and should be specified when possible. Acrylic, however, is a more brittle material.

The same equipment and methods used in machining brass are generally employed when working with plastics.

The above comments are not directly relatable obviously to machining all types of parts with all thermoplastic materials. Deviations are necessary in selecting proper machining methods. Specific recommendations are available on machining data, available through quality local plastic service centers. Plastics definitely lend themselves to punching, sawing, both circular and band, lathe machining, drilling and reaming, milling and routing.

A multitude of mill shape sizes are commercially available to reduce machined part waste and cost.

The materials mentioned above are stock inventory items with quality plastic service centers. Most cities in the United States with a population of 150,000 or more enjoy the services of a reputable plastic mill shape service center.

These centers stock mill shapes in the following configurations:

Sheets	Rods – Round, Octaganol, Hexagonal
Tubes – Round	Tubing – Pressure tubing sizes
Bushing Stock	Bar – Square, Rectangular
Discs	Plates
Cylinders	Billets
Strip	Blocks
Tape	

The diversity of sizes available per shape insures the designer as well as the production foreman of a shape in a material very close to his specification.

The "information gap" between the potential user and the raw materials supplier is attempting to be filled by plastics mill shape processors and distributors. Traditionally plastic mill shape warehousing operations were considered only as sources for materials. Today, they perform an engineering education function. Data sheets are more often seen and used than price lists. A greater appreciation for plastic materials capabilities on the part of the prospective designer has necessitated more product information be available. Materials specification charts, machining information, forming data, use bulletins, design bulletins, and comparative processing bulletins are all available.

Certain sectors of the mill shapes industry has indeed market oriented itself. The needs of the researcher, designer, prototyper, production planner, manufacturing manager and even the mainten-

Table II. Properties of Materials for Heavily Stressed Mechanical

	Cost, Base Resin (¢/in.[3])	*Tensile Strength (1000 p.s.i.)*	*Impact Strength (ft.-lb./in. of notch*	*Abrasion Resistance (mg. loss/1000 cycles)*	*Fatigue Endurance Limit[4] (1000 p.s.i.)*
Nylons[a]	3.7	7.1 to 12.6	0.6 to 4	6 to 8	1.5 3
Acetals[b]	3.3	8.8 to 10	1.2 to 1.4	6 to 20	5
Acetal, TFE-fiber filled[c]	26.5	6.9	0.86	..	..
Polycarbonates[d]	4.5	9 to 10.5	12 to 16	7 to 24	2
Phenolics,[e] fabric filled	1.5 to 3	9 to 16	1 to 2.5	..	..

[a] Vibration damping, nongalling. Available in massive shapes, low friction.

[b] Creep resistant, excellent low-temperature strength, low friction, low moisture absorption.

[c] Self-lubricating, low friction, excellent wear life, creep resistant.

Comments on Table II.

Properties Required: High-tensile plus high-impact strength. Good fatigue resistance and at elevated temperatures. Machinable or moldable to close tolerance.

Suitable Plastics: Nylons, TFE-filled acetals, polycarbonates, and fabric-filled phenolics

Consider Plastics When:

1. Weight reduction is important.
2. Ambient conditions are gritty, abrasive, or corrosive.
3. Part is to be subjected to flexing.
4. Noise or vibration must be controlled
5. Combined functions are desired.

---

ance superintendent are all recognized. This recognition has taken form in perinent product information geared to these levels of interest.

If information concerning the relative merits of forming vs. molding is required, if design factors for sleeve bearings are required,

Components (Gears, Cams, Racks, Couplings, and Rollers)[f]

Flexural Modulus (10^5 p.s.i.)	Heat Distortion Temperature (F) 66 p.s.i.	Heat Distortion Temperature (F) 264 p.s.i.	Resistance to Acids	Resistance to Alkalies	Resistance to Solvents	Resistance to Oils	Machinability
1.5 to 4	340 to 360	140 to 165	F	E	E	E	E
3.7 to 4.1	316 to 338	230 to 255	P	E	E	E	E
4.14	329	212	P	P	E	E	E
3.2 to 3.8	283 to 293	270 to 280	E	G	G	F	E
8 to 14	320	320	F	F	E	E	F to E

E=Excellent G=Good F=Fair P=Poor

[d]Extremely creep resistant, transparent, excellent low temperature strength, low moisture absorption high dimensional stability.

[e]Hard, extremely creep resistant.

[f]This table was taken from *Plastics Quarterly, Machine Design.* (Dec. 12, 1968).

Consider Other Materials When:

1. Low inertia and high starting speeds are required.
2. Cost is all important.
3. Loadings are heavy.
4. Service temperatures are high.

Consider Plastic-Other Materials When:

1. Low cost and flex resistance are required.
2. High impact resistance is required.

Property Summary: Nylons are recommended for general-purpose gears and other mechanical components. Acetals for maximum fatigue life, for highly accurate parts, or exposure to extremely humid conditions. Phenolic-fabric laminates for low-cost, thin stamped gears or parts. Polycarbonates for intermittent, very high impacts (not recommended for applications involving repeated cyclical stress). TFE-filled acetals for heavy-duty applications.

---

if specific machining information of plastics is required, this and more is available.

Table III. Properties of Materials for

	Cost, Base Resin (¢/in³)	*Tensile Strength (1000 psi)*	*Impact Strength (ft-lb/in. of notch)*	*Brittle Point (F)*	*Heat Resistance Continuous (F)*	
					Range	*Typical*
Fluorocarbons[a]						
TFE & FEP	25 to 43	1.5 to 4.5	2.5 to 16	-420	400 to 550	
CTFE[b]	45	4.6 to 5.6	3.1 to 7.3	-400		400
Chlorinated[c] Polyether	12.6	6	0.4	10 to -20		290
Polyvinylidene[d] Fluoride	37.5	7		<-80		300
Polypropylenes[e]	.09 to	3.3 to	0.3 to	0	230 to	275
Polyethylenes,[f] high-density	0.9	2.9 to 5.5	0.4 to 14	-76 to -200	170 to 260	250
Polyamides[g]		10.5	0.9		500 to 900	500
Polyphebylene[h] Oxide	4.4	11.6	1.3		250	
Epoxy-Glass[i]	3.4 to 6.5	34 to 100	10 to 24		250 to 400	250 to 300

[a]Has good mechanical properties over full temperature range, low friction zero moisture absorption. FEP higher priced, but may be injection molded.

[b]Transparent, injection moldable, zero moisture absorption, resistant to radiation and creep.

[c]Good abrasion resistance.

[d]Injection moldable and extrudable.

[e]Lightest of plastics; excellent resistance to creep and stress cracking.

Comments on Table III.

Properties Required: Resistance to temperature extremes and to wide range chemicals. Minimum moisture absorption. Fair to good strength.

Suitable Plastics: Fluorocarbons, chlorinated polyether, polyvinylidene fluoride, polypropelene, high-density polyethylene, and epoxy glass.

Chemical and Thermal Equipment [j]

Heat Distortion Temp. at 66 p.s.i. Range	Typical	Flammability	Effect of Strong Acids	Strong Alkalies	Solvents
160 to 260	250	None	None	None	None
196 to 391	265	None	None	None	Very Slight
	300	Self-ext.	Very Slight	None	Very Slight
	300	Self-ext.	Very Slight	None	Very Slight
215 to	210	Slow	Very Slight	Very Slight	Slight
140 to 180	175	Very slow	Slight	Very Slight	Slight
	>470	non-burning	Attack	Attack	Attack
	355	Self-ext. non-drip	Very Slight	Very Slight	None
300 to 550	350 to 375	Slow to None	Slight	Slight	None

[f]Good abrasion resistance; lighter than water.

[g]Good mechanical and physical properties in a wide temperature range; excellent ionizing radiation resistance.

[h]Good overall properties even at high temperatures.

[i]Readily applied by hand layup to large areas, adheres tightly to most substances.

[j]This table was taken from *Plastics Quarterly, Machine Design* (Dec. 12, 1968).

Consider Plastics When:

1. Cost is a primary consideration.
2. Ultimate in corrosion resistance is required.
3. Abrasives may be present in combination with corrosives.
4. Minimum maintenance is desired.
5. Thermal insulation is a requirement.

Comments on Table III (Continued).

Consider other Materials When:

1. Maximum strength properties are needed.
2. Service temperatures exceed 550°F.
3. Dimensional stability must be good over wide or fluctuating temperature ranges.

Consider Plastics-Other-Material Combinations When:

1. Optimum strength properties and maximum corrosion resistance are required.
2. Intense heats are involved and the plastic ablates slowly, preventing heat from damaging the metal.

Property Summary: TFE fluorocarbons for general-purpose chemical and extreme temperature applications. Polypropylene and high-density polyethylene for plating and less severe chemical exposures. Chlorinated polyether, PVF and CTFE fluorocarbons for extreme resistance in combination with mechanical strength and stiffness. CTFE for transparency. Epoxy-glass for greatest mechanical strength and for large structures.

Mechanical Plastics Pricing History and Structure

Current mill shapes pricing falls into three categories:

End-User
Distributor
O.E.M.

For our discussion, we will concern ourselves directly with end-user pricing and distributor and original equipment manufacturer as variations of end-user price levels. The following end-users price levels reflect current market costs per material in 1 inch thick slab sections. They are compared with pricing levels of 1958.

Current		*1958*	
Fluorocarbons	$4.50 per pound	Fluorocarbons	$8.64 per pound
Nylon	2.50 per pound	Nylon	2.90 per pound
Acetal	2.70 per pound	Acetal (1963)	3.00 per pound
Polycarbonate	3.00 per pound	Polycarbonate (1963)	3.00 per pound
Polyolefins	.90 per pound	Polyolefins	1.45 per pound

This is because of lower raw materials costs through increased production on the part of the resin manufacturers as well as continuing manufacturing efficiencies effected by the mill shape processors.

These costs declines coupled with plastic materials lower specific gravity allow them to compete more effectively on a price basis with metals in many instances.

Distributor pricing of mill shapes represents a discount from the above prices, excepting fluorocarbons, of approximately 30%. Of course, large (O.E.M.) original equipment manufacturer special runs affect a further discount which is variable depending on the size, type, and quantity of the product involved.

Machining costs of plastics can be discussed only in general terms unless specific jobs are examined. It will suffice to say that $10.00 per hour is a rule of thumb. Obviously a Swiss Automatic can command because of higher output perhaps $30 to $40 per hour. Milling and lathe turning can be lower than the $10.00 per hour figure. A reasonable average figure would therefore be $10.00.

The sales and distribution channels of plastic mill shapes are less sophisticated and not as structured as distribution of ferrous and non-ferrous metals. The industry growth has been so rapid that maturity of distribution as found in other mill shapes industries has not had time to evolve. Currently there are almost twenty mill shape producers of mechanical plastics. These producers supply the market through distributors, direct to the O.E.M. and in one instance a national producer employs the above marketing modes plus utilizing its own fifty stocking service centers. This places the company in the closest possible contact geographically with all its markets.

Plastic shape distributors number about 317 throughout the Continental United States. Most carry modest inventories and are single unit locations. Parts or component fabrication usually bolsters their business effort. In addition to these 300 plus plastic distributors, approximately 2,000 other industrial distribution firms round out their lines or service goods customer requirements with an occasional length of rod or length of plate. These peripheral operators are really not in the business but offer product as a courtesy item. Many industrial supply houses, bearing supply houses, adhesive companies, and chemical supply houses, make up this category of subdistributor classification. For the most part, they are not materials oriented and can offer little concrete data or application information; and therefore, offer only a supply service for usually a very limited variety of products. These products are rarely inventoried but purchased from a local plastics service center.

The System Approach to Product Development

It is important to note that some manufacturers of mill shapes in order to cope with continuing need for greater in-depth material and applications needs of potential customers have augmented the sales effort of their branch service centers with centrally located product

Table IV. Properties of Plastics for Light-

	Cost, Base Resin (¢/in.[3]	*Tensile Strength (1000 p.s.i.)*	*Impact Strength (ft-lb/in. of notch)*	*Flexural Modulus (10^5 p.s.i.)*	*Heat Resistance Continuous (F)*
Acrylics[a]	2.2	5.5 to 10.5	0.4 to 0.5	3.5 to 5	150 to 225
Polystyrenes[b]	0.6	5 to 9	0.25 to 0.4	4 to 5	150 to 175
Acetates[c]	1.8 to 2.4	1.9 to 11	1.2 to 5.8	1.1 to 4	180 to 200
Cellulose[d] Acetate Butyrates	2.7 to 3.3	2.6 to 6.8	0.8 to 6.3	0.6 to 1.8	140 to 220
Vinyls, Rigid[e]	1.2 to 2.1	5.5 to 9	0.25 to .12	3.8 to 5.4	150 to 220
Polycarbonates[f]	4.5	9 to 10.5	12 to 16	3.2 to 3.8	250 to 270
Styrenes,[g] medium, impact	0.7 to 1.1	3.5 to 6.8	0.6 to 3	3 to 7.5	155 to 180
Ionomers[h]	16.7	5	5 to 14	0.4	160

[a]Highest reflectance (sparkles in reflected light) of transparent plastics, pipes light excellent low temperature properties.

[b]High reflectance, pipes light, excellent low temperature properties, brittle, machines poorly stress-crazes.

[c]Hand formable.

[d]Takes extremely deep draws.

Comments on Table IV.

Properties Required: Good light transmission in transparent or translucent colors. Good to excellent formability and modability. Shatter resistance. Fair to good tensile strength.

Suitable Plastics: Acrylics, polystyrenes, cellulose acetates, cellulose buyrates, ionomers, rigid vinyls, polycarbonates, and medium-impact styrenes.

Other Suitable Materials: Glass.

Consider Plastics When:

1. Shatter resistance is required.
2. Virbration resistance is important.
3. Flexibility is required.

Transmission Components, Glazing Models [j]

Light Transmission (%)		Haze (%)		Effect of Ultra-	Form-	Resistance to			
Range	Typical	Range	Typical	violet	ability	Acids	Alka-lies	Solvents	Oils
91 to 93	92	1 to 3	1	None	G to E	G	F	F	F
75 to 93	..	>3	..	Slight to Crazes	P	G	E	P	E
75 to 95	80	2 to 15	9	Slight	G	P	P	F	G
80 to 92	88	1 to 4		Slight to None	P	P	P	G	E
..	80	3 to 4	4	Slight	G to E	G	E	G	G
75 to 85	80	>10		Fade	E	E	G	G	F
10 to 55	30			Slight	G	G	E	G	G
	95		3			G	E	G	E

[e]Good abrasion resistance; excellent dielectric; printable.

[f]High creep resistance and creep stability.

[g]Translucent only.

[i]Tough; excellent clarity.

[j]This table was taken from *Plastics Quarterly, Machine Design* (Dec. 12, 1968).

4. Colored transparency is desired.
5. Maximum strength-to-weight ratios are required.
6. Translucency must be obtained within the material rather than by surface treatment.
7. Ease of forming in complex shapes is required.
8. Hand fabrication of prototypes is required.

Consider Other Materials When:

1. Maximum chemical resistance is required.
2. Highly abrasive conditions are present.
3. High temperatures are encountered.

Comments on Table IV (Continued)

4. Maximum dimensional stability over wide range of temperatures is needed.
5. Absolute imperviousness to moisture is required.

Property Summary: Acrylics recommended for general-purpose applications, especially for optical decorative, and outdoor use. In sheet stock, cast acrylic has greater strength and transparency extruded acrylic costs less (especially in thin members), and has better formability. Polycarbonates for maximum strength, as in explosion shields. Butyrates for excellent impact resistance, and deep formability. Vinyls for maximum formability and printability. Acetates and vinyls for flexible glazing and guards. Medium-impact styrene and rigid vinyls for lowest-cost molded transparent parts.

---

manager groups. The individual product managers have product development as well as market development responsibilities.

This additional marketing service has allowed a much greater in-depth penetration of plastic material capabilities. Such service is available through less than a handful of plastic mill shape processors.

Approximately 18 months ago, one such product manager while interviewing members of the brewing industry, noticed a wear pattern appearing in several areas around bottle filling equipment and bottle labeling equipment. One wear problem existed under the infeed and outfeed conveyor chain. Currently, oil impregnated wood, brass and bronze had been used. Acetal molded parts laping tongue in groove as well as custom extruded molybdenum disulfide filled nylon were common to the application. While each product offered one or two specific advantages, only ultra-high molecular weight polyethylene as specified and designed by the product manager fulfilled all the requirements of:

1. Long wear life.
2. Extremely low co-efficient of friction.
3. Moisture resistance.
4. Long length availability for ease of installation.

This section is fabricated from a standard mill shape.

Another excessive wear problem was occurring in star wheels, sometimes called spiders or indexing stars. These parts serve a function of spacing and feeding bottles or containers in and out of filling, labeling and inspection stations on conveyor lines. Since the parts involved are indexed at up to 800 per minute, any malfunction causes tremendous product (beer) loss and container breakage. As glass bottles are extraordinarly abrasive a wear resistant and resilient material is required which is resistant to moisture attack from in-place stream cleaning.

Brass, bronze, steel, phenolic, acetal, and nylon (natural) had been used with various degrees of success. A specially formulated cast polyamide was recommended, tested, and designed into the function. The part is currently offering a longer wear life, with less moisture pick-up and greater ease of installation.

The part is sold as a fabricated part to the end-using brewery. However, the mill shape itself is available if the brewery personnel elect to machine the part themselves.

These two instances represent related areas where specific properties of plastic mill shapes properly evaluated and correctly designed can extend part wear life, reduce installation cost and reduce handling, and create greater production efficiencies, at perhaps the same or a lower cost.

Plastics have good combinations of properties rather than extremes of any single property. For example, no plastic approaches the steels in strength. No plastic is as light in solid form as most woods, as elastic as soft rubbers, as scratch-resistant or transparent as glass. Yet plastics are the only materials which can be simultaneously strong, light weight, flexible, and transparent.

Similarly, nylon and fluorocarbon are outstanding dry bearing materials not merely because of their low coefficient of friction but because they offer in addition, a combination of good abrasion and corrosion resistance, fair to good creep resistance, and the ability to absorb abrasive particles. In some instances mechanical plastics are used because of one outstanding property—fluorocarbon for anti-stick characteristics, for example—but these are not typical.

Plastic parts require a minimum of finishing after processing. They can have integral color, gloss, or textured finish imparted during molding or forming, and threads undercuts, and metal inserts or holes molded in. A single plastic molding may replace an assembly of several metal components.

Because of their combination of properties, plastic parts may provide several functions. They may be both mechanism and housing; integral shaft, bearing, or gear (or all three) integral container, housing, and insulation.

The choice of a particular plastic includes numerous types. It is easier to cope with this multitude of product if plastics are considered as a category of materials—like metals—encompassing some 40 plus distinct families. Each family in turn is composed of many individual types. Just as the family of steels includes carbon steels, tool steels, stainless, etc., nylon includes 6, 6/6, 6/10, filled or unfilled, etc., each type having distinct properties.

While there are a great number of plastic materials with which to design a certain function in practice any given application usually lies within a narrow band of materials.

Within this band, several plastic materials may be suitable. The final decision involves a close comparison of the various properties offered by the materials under consideration.

Evolving from the one hundred year existence of the plastics, mill shapes have taken their place as engineering products adequate to the needs of the engineering and design communities. Improved properties in new materials combined with processing and machining techniques, have brought "to age" mechanical plastic mill shapes as engineering materials. Creative product development methods combined with sophisticated marketing exposure, have established this "industry" within an industry.

Literature Cited

(1) Dourlet, Ernest F., *Plastics,* Quarterly, Machine Design, p. 8 (Dec. 12, 1968).

RECEIVED, June 6, 1969.

12

Thermoplastics from the User's Standpoint

J. M. CAMERON[1]

Fisher Body Division, General Motors Corp., Warren, Mich. 48093

The automobile industry has become a significant contributor to the tremendous growth of the plastics industry in the past five years. Total plastic consumption in the automotive market is expected to continue the phenomenal growth through at least the next ten years. The variety of plastics is continuing to expand, floor space devoted to plastics is increasing, and processing methods are becoming more sophisticated. The mutual cooperation of both the marketing people and the fabricators are required to insure the continued growth of thermoplastics. The flow of information, which is necessary, can be accomplished through personal contact with design engineers, material engineers, product engineers, and research and development engineers.

The automobile industry has become a significant contributor to the tremendous growth of the plastics industry in the past five years. In 1965 there was approximately 30 pounds of plastic per automobile. In 1969 there is approximately 85 pounds of plastic per automobile. The total plastic consumption by the automotive industry is expected to continue this phenomenal growth through at least the next ten years.

According to a survey conducted by the Society of the Plastics Industry, among the 360 exhibitors at the 12th National Plastics Exposition, the automotive industry was placed in the No. 1 position as having the greatest growth potential of any major market.

The Fisher Body Division of General Motors is the largest consumer of plastic automotive parts and one of the largest consumers of plastic raw materials in the United States. At the present time, our organization is purchasing ten generic families of thermoplastic resins consisting of 35 various types.

[1] Present address: Swedish Crucible Steel Co., Detroit, Mich. 48211

Generic Families of Thermoplastics Used By Fisher Body Division of General Motors

1. Acrylonitrile-Butadine-Styrene (ABS)
2. Cellulose Acetate Butyrate (CAB)
3. Ethylene Ethyl Acrylate (EEA)
4. Polyamide (Nylon)
5. Polyethylene (PE)
6. Polymethyl Methacrylate (Acrylic)
7. Polypropylene (PP)
8. Polystyrene (PS)
9. Poly (vinyl chloride) (PVC)
10. PVC/ABS Blends

The manufacturing plants which produce thermoplastic components are located in Detroit, Michigan, Cleveland and Elyria, Ohio, Syracuse, New York, and Trenton, New Jersey.

Plastic production and development floor area being utilized is approximately 625,000 square feet. This is an increase of 50% over the previous production year. The total pounds of thermoplastics which will be converted during the 1969 model year will be approximately five times the amount converted in 1965.

The production facilities utilize the following manufacturing processes:

1. Blow Molding
2. Extrusion
3. Injection Molding
4. Casting of Plastisols
5. Vacuum Forming

The research and development activities of the Division are located at the General Motors Technical Center in Warren, Michigan. At this location, the ideas and concepts of today are created and developed into the products of tomorrow.

Since this paper is primarily concerned with working with the market researcher and the market development man, I should define my interpretation of their responsibilities.

The market researcher is one who researches a particular market to determine if the potential is great enough to warrant development of a new product or perhaps locate markets for a product in the development stages.

The market development man is responsible for the development of markets for new or existing products and to expand the usage of his product in current markets.

To assist the market research and market development people in their endeavor to locate and develop potential automotive markets for their products, I would like to describe the various stages of product development to give you a better appreciation of our organization.

The initial concept for a component may be initiated by either product design or new product development. At this time the initial concept is primarily a guide for the final detailed design.

The operating parameters of the particular part and the material requirements are outlined to determine whether the concept is limited by materials or processes. Usually, on the basis of the broad requirement analysis, certain theromoplastic materials can be eliminated and several logical candidates can be selected.

If it is determined that material limitations are an obstacle, modification of the concept may be possible. On the other hand, a material development or modification program may be initiated to obtain a material that will meet the requirements of the concept.

The actual selection of the material is performed by the material engineering group in conjunction with product design or new product development. Also, the product engineer can furnish additional information relative to part location, mechanical and physical requirements. Thus, the various design factors, product requirements and material requirements are thoroughly analyzed early in the design stages.

In the engineering development stage, a practical and workable design is developed. Prototype samples are produced on equipment located at the Technical Center that duplicates the production process to be used. A portion of the materials evaluation work is also performed at this time. Candidate materials are given performance tests and moldability is analyzed in detail. Cost and reliability factors are also considered at this time.

Upon completion of the preliminary evaluation of design and materials, any required revisions are made. The next step is to fabricate several hundred parts from the material which performed most satisfactorily during the initial evaluation. These parts will be subjected to many types of physical and mechanical tests in the laboratory. Parts are also installed in automobiles at the G.M. Proving Grounds and fleet cars to obtain actual field experience.

The outcome of the development stages are the final production drawings which are released to the manufacturing plant for production. An interim step is the ordering of tooling as soon as the production drawings are available to tool engineering. Once the tools and equipment are available to production, a tool and material tryout is conducted and any problems that may exist are corrected.

Interest in a part does not terminate once it is in production. The evaluation of part performance, materials, and processing are periodically reviewed. We may take advantage of a new material to improve processing, reduce production costs or reduce material costs.

Working with an Approved Supplier's Market Researcher

The market researcher may find his job of calling on the automotive accounts more difficult than smaller, less complex industries. Because of the size and complexity of automotive divisions, a single individual is generally not able to supply sufficient information to complete a market survey of one division. One must contact numerous people in various areas to obtain the necessary information to evaluate the market potential.

We are continually looking for new and improved materials to reduce cost and/or improve our products. We maintain an "open-door" policy and are willing to discuss our present and future needs with the market researcher

The purchasing department is willing to discuss present requirements, future requirements, and approximate volumes of materials. Purchasing is also able to give guidance as to which engineering group would be involved and the engineer to contact.

The product engineering department is divided into groups depending on location of the part in the body and its function. Generally, the product engineer can supply part requirements, approximate volumes, and projected usage.

Material engineers can discuss the material requirements of current production parts and give an indication of possible applications for a new or modified polymer.

Working with Market Development People

We rely on the market development people to help keep us up to date on the latest material developments and keep their research and development people informed of our present and future material requirements.

There is a co-operative effort between our organization and the material suppliers relative to material development and evaluation programs of new or modified polymers. However, the majority of application development work is conducted within the division.

To proceed from the contacts completed by the market research people, the market development man will find many more contacts are required.

It is necessary for all suppliers to arrange their visits through the purchasing department. The purchasing department will assist in directing suppliers to the engineers involved in the various aspects of a particular program. This may entail contacting design engineers, project engineers, process engineers, material engineers, and production engineers. If all of the engineers involved in a particular program are contacted, the market development man would be insured of keeping all lose ends together and staying abreast of the program.

When the market development man discusses his material with the engineers, he should be prepared to discuss the characteristics that are most important to them.

The properties of thermoplastics used most frequently by our engineers are as follows:

1. Specific Gravity
2. Hardness
3. Mold Shrinkage
4. Water Absorption
5. Outdoor Weathering
6. Coefficient of Linear Thermal Expansion
7. Tensile Strength
8. Elongation
9. Flexural Yield Strength
10. Flexural Modulus
11. Izod Impact
12. Deflection Temperature @ 264 & 66 p.s.i.
13. Shear Strength
14. Compressive Strength

In addition to the above, it is helpful to have information relative to processing, decorating, bonding, and of course, cost.

Mechanical Property Data

It would be helpful to our engineers in making material comparisons if direct comparisons of properties could be made from one thermoplastic to another.

One problem with some of the technical data sheets on thermoplastic compounds, is lack of uniformity. The variations which exists between suppliers are in the size of the test specimens, the speed at which the particular test is performed, and in some cases the temperatures in which the test is performed. To accomplish increased uniformity, it is suggested that the size of test specimens be standardized for a particular test and not be varied to display a more impressive number. Test speeds should be established for the test and not for a particular type of thermoplastic. Also, standardization of temperatures to be used if data are reported at other than 73°F.

Sales and Service Requirements

The sales and service requirements of the material supplier to the automotive industry are very demanding. The pace is rapid and our continuing efforts to increase quality, improve product reliability, and give our customers the most for his dollar are some of our continuing goals.

The purchasing department generally gives consideration to any supplier who makes a contribution towards improving our products. However, the buyer has an obligation to obtain competitive quotations except under unusual and justifiable circumstances. This cannot be accomplished without the aid of competitive bidding. This is not to say that the lowest bidder is always awarded the business, it is dependent on total economics.

In addition to buying a reliable material at a fair price, the buyer tries to locate sources that channel a portion of their income into research and development which invariably leads to improved materials and better ways of making products.

To be most effective, it is helpful for suppliers to have sales offices located in the Detroit area. Trying to discuss technical problems and other engineering business by telephone or letter is rather cumbersome and generally not the most efficient method. There is sufficient activity in the Technical Center alone to warrant assigning a market development man and/or salesman on a full-time basis.

Summary

It has been demonstrated that the roles of the market researcher and the market development men are important not only to the growth of the material suppliers, but also to the growth of the fabricators.

The mutual cooperation of both the marketing people and fabricators is required to insure the continued growth of thermoplastics. The flow of information, which is necessary, can be accomplished through personal contact and purchasing, material engineers, product engineers, and research and development engineers.

RECEIVED June 6, 1969. The information presented in this paper is pertinent to the Ternstedt Division of General Motors prior to the merger with the Fisher Body Division. The effective date of the merger was November 1, 1968.

INDEX

INDEX

N

O

P

R

S

T

U

V